MATH SKILLS FOR THE SC

MATH SKILLS
FOR THE SCIENCES

JOHN G. PEARSON
University of Alaska

DENNIS M. STONE
University of Michigan

RICHARD F. SWINDELL
Mt. Growatt College, Australia

under the auspices of
The University of Idaho
and
The Idaho Research Foundation, Inc.

John Wiley & Sons, Inc.
New York • London • Sydney • Toronto

Editors: Judy Wilson and Irene Brownstone
Production Manager: Ken Burke
Editorial Supervisor: Winn Kalmon
Composition and Make-up: Shoshana Weinman

Library of Congress Cataloging in Publication Data

Pearson, John G
   Math Skills for the sciences.

   (Wiley self-teaching guides)
   "Under the auspices of University of Idaho and the Idaho Research Foundation, Inc."
   First published (© 1972) under title: Math skills for first year science.
   Includes index.
   1. Mathematics—1961-   I. Stone, Dennis M., joint author. II. Swindell, Richard F., joint author. III. Title.
QA39.2.P4   1976   510   75-40065
ISBN 0-471-67541-5

Printed in United States of America
76 77 10 9 8 7 6 5 4 3 2

ACKNOWLEDGMENTS

We wish to thank the Associated Students of the University of Idaho, and the Idaho Research Foundation for financial support that culminated in the publication of our original book under the title of *Math Skills for First Year Science*. Our thanks also to R. W. Stark of the Idaho Research Foundation and to Mike Hardie of the University of Idaho who revised the book into its present form for a wider audience.

Additionally we wish to thank our wives for their understanding, patience, and beer money.

To the Reader

This book is not meant to be a scholarly text. Any math purist will agree that if we intended to write a non-scholarly work, we have succeeded.

We have written this book because we have noticed that science students frequently have more trouble with the arithmetic associated with problems than with theory and concepts. In fact, it is our experience that those complaints that chemistry, physics, and biology are difficult subjects are most heard during those parts of the courses where mathematics is required. The source of the difficulties has nothing to do with a student's intelligence, but rather with rusty skills or gaps in his educational background. Many science teachers feel a corresponding frustration (and rightly so) that they are teaching too little science and too much mathematics.

For these reasons we have written a book which can teach math skills for the sciences without the supervision of an instructor. The book was constructed with all the meaningful manipulations but with none of the nasty notations which obscure the obvious.

If you wish to brush up your math skills, you can use this book on your own. Each chapter begins with a pretest. You need only study those parts of the text that correspond to pretest questions you cannot answer. Therefore, you should take the pretest and then check the answer key, which tells you what part of the chapter (if any) you should study. Each pretest question corresponds to a specific topic within the chapter, so that you can study examples and work problems until you have mastered the skills in that section. After completing a particular chapter, you can review the pretest to make sure you can do *all* the problems. Then go on to take the pretest for the next chapter.

Each chapter is divided into short numbered sections called frames. Each frame covers new material and provides practice problems. Answers for each frame are separated from the problems by a line of dashes. You should cover the answer section and write your own answer before checking the book's solution. If your answer is incorrect, be sure you find out where you went wrong and correct your answer before you go on. You'll learn best if you actually work out all the problems yourself. In this way math skills will become automatic and problem solving easy.

In the early part of this book and where problems require a simple answer, we have left space for you to write your computations directly in the book. You may, however, want to keep a pad of paper handy for more complicated calculations. Later in the book, you will do your problem solving on a separate sheet of paper.

At the end of the book is a Final Self-Test so you can evaluate your learning. The test is subdivided by chapters, and the answers, also subdivided, follow the text. If any of these test questions give you difficulty, return to the pretests and review the appropriate material.

We would like to point out that when the material in this text was studied by students with low math aptitudes, they averaged a whole letter grade higher in chemistry than did their classmates with the same math aptitude who did not study the material.

February 1976

John G. Pearson
Dennis M. Stone
Richard F. Swindell

Contents

Chapter One	GENERAL MATH OPERATIONS	1
Chapter Two	MATH OPERATIONS ON FRACTIONS	15
Chapter Three	ALGEBRA	29
Chapter Four	EXPONENTS	51
Chapter Five	SCIENTIFIC NOTATION	68
Chapter Six	LOGARITHMS	77
Chapter Seven	PROBLEM SOLVING AND DIMENSIONAL ANALYSIS	105

FINAL SELF-TEST — 136

LOGARITHM TABLES — 143

INDEX — 146

MATH SKILLS FOR THE SCIENCES

CHAPTER ONE
General Math Operations

PRETEST AND OBJECTIVES

This pretest outlines the objectives for Chapter One. It should help you
identify what parts of the chapter, if any, you need to read. Try each of the
problems which follow. Then check your answers with the answer key, and
follow the directions given there.

Objectives and Pretest   **Your Answer**

When you complete this chapter, you will be
able to:

1. Apply the correct sign when adding or subtracting integers or expressions.

   (a) $4 + 8 =$ _____

   (b) $6 + (-7) =$ _____

   (c) $(-8) + (-4) =$ _____

   (d) $3a - (-5a) =$ _____

   (e) $(-12) - (+4) =$ _____

   (f) $6a + 3b =$ _____

2. Recognize and apply standard multiplication
   terms and notation.

   (a) In the equation, $3x = 6$, which terms
        are factors?   _____

   (b) Identify the product in the equation,
        $2 \times 4 = 8$ _____

   (c) Simplify $5(a + b)$ _____

   (d) Simplify $2(a \cdot b)$ _____

   (e) $92 \cdot 0 =$ _____

(f)   What is the product of $6 \cdot x \cdot y \cdot 2$?   _____

3. Apply the correct sign when multiplying integers or expressions.

(a) $3 \times (-3) =$ _____

(b) $-A(B - 1) =$ _____

(c) $(-4)(-2)(-1) =$ _____

(d) $(3)(-5)(-a) =$ _____

4. Recognize and apply standard division terminology and notation.

(a)   Perform the indicated operation
       $12 \div 3 =$ _____

(b)   Perform the indicated operation
       $3\overline{)9} =$ _____

(c)   If 5 is the divisor and 15 is the dividend, what is the quotient?   _____

(d)   The numerator of the fraction $\dfrac{X}{B}$ is   _____

(e)   The ratio of 4 to 5 would be written as   _____

(f)   The reciprocal of $\dfrac{C}{D}$ is   _____

(g)   The inverse of 2 is   _____

(h)   Express $\frac{4}{8}$ as a decimal fraction.   _____

(i)   Given $\dfrac{9b}{3c}$, how many b per c?   _____

(j)   Simplify the expression $\dfrac{4ab(Q + b)}{2b}$ _____

5. Apply the correct sign when dividing integers or expressions.

(a) $\dfrac{-12}{-3} =$ _____

(b) $\dfrac{-4}{-2} =$ _____

(c) $\dfrac{6a}{-3} =$ _____

(d) $\dfrac{(20ab)(-2)}{-5} =$ _____

Answer Key and Directions

Check your answers with the answer key which follows. If you missed any of these questions or if you want help with a particular objective, read the indicated portion of Chapter One.

If you missed any of these problems: **See frames:**

1. (a) 12
   (b) −1
   (c) −12
   (d) 8a
   (e) −16
   (f) 6a + 3b   frame 1, page 4

2. (a) 3 and x
   (b) 8
   (c) 5a + 5b
   (d) 2ab
   (e) 0
   (f) 12xy   frame 2, page 6

3. (a) −9
   (b) −AB + A
   (c) −8
   (d) 15a   frame 3, page 7

4. (a) 4
   (b) 3
   (c) 3
   (d) X
   (e) $\frac{4}{5}$
   (f) $\frac{D}{C}$
   (g) $\frac{1}{2}$
   (h) .5
   (i) $\frac{3b}{1c}$ or 3b/c or 3b per 1c
   (j) 2a(Q + b) or 2aQ + 2ab   frame 4, page 8

5. (a) 4
   (b) +2
   (c) −2a
   (d) 8ab   frame 5, page 13

If you worked all problems easily and correctly, go on to the Chapter Two Pretest and Objectives.

POSITIVE AND NEGATIVE SIGNS

(1)   Many of the scales, instruments, and quantities used in science have plus and minus values. The sign that precedes a number tells us whether the quantity is larger or smaller than zero. If the sign is positive, +, the quantity is larger than zero. If the sign is negative, −, the quantity is smaller than zero. The number or symbol following the sign tells us the magnitude of the largeness or smallness.

    Examples: +5 is five units greater than zero.
                   +5a is five units greater than zero a's.
                   −5 is five units smaller than zero.
                   −10 is ten units smaller than zero and is five units
                      smaller than −5.
                   −60 degrees is colder than zero degrees.
                   −70 degrees is a lower temperature than −60 degrees.

To differentiate between math operations (such as addition or subtraction) and the sign of a quantity, we place the quantity and its sign within parentheses. Parentheses need not be used if the quantity is not being used in a calculation.

    Examples: −2·x −2 should be (−2) x (−2).
                   +4 + −2 should be (+4) + (−2).

Generally, if a number is greater than zero the plus sign is understood and not expressed. If the sign of a quantity is not indicated, the quantity is positive and there is no need to include parentheses.

    Examples: (+4) x (+3) is the same as 4 x 3 and either expression is
                      correct.
                   (+4) + (+3) is the same as 4 + 3.
                   (+4) + (−3) is the same as 4 + (−3).
                   (+6) − (+4) is the same as 6 − 4.

ADDITION AND SUBTRACTION RULES

Rule 1. To add quantities with the same sign, add and keep the sign.

    Examples: 4 + 2 = +6 or 6
                   a + 3a = +4a or 4a
                   (−5) + (−6) = −11
                   (−a) + (−3a) = −4a

Rule 2. To add quantities with different signs, first drop the signs. Then subtract the smaller quantity from the larger. Finally, attach the original sign of the larger quantity.

Examples: $(-3) + 5 = +2$ or 2
$(-5) + 3 = -2$
$(-a) + 2a = a$
$(-3a) + 2a = -a$
$6 + (-4) = 2$

Rule 3. To subtract one quantity from another, change the sign of the quantity to be subtracted and add the two quantities, following Rule 1 or Rule 2.

Examples: $4 - (+2) = 4 + (-2) = +2$
$5 - (+8) = 5 + (-8) = -3$
note: $4 - (+2) = 4 - 2 = 2$
$5 - (+8) = 5 - 8 = -3$
$(-7) - (+2) = (-7) + (-2) = -9$
$(-6) - (-3) = (-6) + (+3) = -3$
$(-2a) - (+5a) = (-2a) + (-5a) = -7a$

Rule 4. The addition or subtraction of unlike quantities is not permitted.

It is impossible to add apples and oranges. (That is, 3 oranges + 2 apples does not compute!) We can only say that three oranges and two apples are equal to three oranges and two apples. If we were considering fruit in general, we could add the number of fruit, 3 fruit (oranges) + 2 fruit (apples) = 5 fruit (apples + oranges). However, we cannot add unlike symbols.

Examples: $5a + 7a = 12a$
$7a - 5a = 2a$
$6a + 4b = 6a + 4b$ (not $10ab$)
4 quarts + 2 quarts = 6 quarts
4 quarts + 2 gallons = 4 quarts + 2 gallons
(Because 2 gallons = 8 quarts we can calculate the result, 4 quarts + 8 quarts = 12 quarts, but we had to change the meaning of the quantity, gallons.)

PROBLEMS. Carry out the indicated operations.

(a) $3 + (+6) =$ _____   (e) $(-3a) - (+6a) =$ _____

(b) $6 + (-3) =$ _____   (f) $(-8) - (-4) =$ _____

(c) $(-7) + 4 =$ _____   (g) $(-2a) - (-6a) =$ _____

(d) $4 - (+2) =$ _____   (h) $3y - (-4z) =$ _____

– – – – – – – – – – – – – – – – –

(a) $3 + (+6) = +9$ (e) $(-3a) - (+6a) = -9a$
(b) $6 + (-3) = +3$ (f) $(-8) - (-4) = -4$
(c) $(-7) + 4 = -3$ (g) $(-2a) - (-6a) = 4a$
(d) $4 - (+2) = +2$ (h) $3y - (-4z) = 3y + 4z$

If you did any of these incorrectly, make sure you find out where you went wrong. Correct your answer before you go on.

MULTIPLICATION TERMINOLOGY AND NOTATION

(2)   Most of us are quite familiar with the process of multiplication and how it is related to addition (for example, 3 x 3 is the same as 3 + 3 +3). However, some of us may be unfamiliar with important terminology and varied means of representing multiplication. These are reviewed below.

factor
A quantity that is multiplied by another quantity; may also be called a coefficient.

Examples: 2 x 3 = 6 (2 and 3 are factors)
            a x b = ab (a and b are factors)

product
The result of a multiplication operation.

Examples: 2 x 3 = 6 (6 is a product)
            4 x 12 = 48 (48 is a product)

zero factor
Any quantity multiplied by zero is zero.

Examples: 1024 x 0 = 0
            $9ab$ x 0 = 0

x and ·
An x between two quantities means multiply the two quantities. A dot between two quantities also means multiply the two quantities.

Examples: 2 · 3 = 2 x 3 = 6
            2 · a = 2a
            4 · 10 = 40

Often the dot or x is understood and not written. This presentation is used when:

(1) At least one of the factors is a symbol.

Examples: 2 x b = 2b
            a x b = ab
                Note: 4 x 2 does not equal 42

(2) One of the factors is written within parentheses. Parentheses are placed around a sum of quantities, or quantities with negative signs.

Examples: $5 \times (-3) = 5(-3) = -15$
$(-3) \times (-2) = (-3)(-2) = 6$
$A \times (-2) = A(-2) = -2A$
$5 \times (a + b) = 5(a + b) = 5a + 5b$

> Note: the factor 5 is multiplied with each term within the parentheses.

The process shown in the last example above is called *distributing multiplication over addition.* You must use this process every time you multiply a quantity with a sum or difference inside parentheses. If the quantity inside the parentheses is not a sum or difference, you only need to multiply once. *Simplify, Distribute,* or *Multiply Out* are common directions accompanying this process.

Examples: $5 \times (8 - ab) = 40 - 5ab$
$5 \times (8ab) = 40ab$

PROBLEMS.

(a) In the equation $7c = 42$, which terms are factors? _____

(b) Identify the product in the equation, $3 \times 8 = 24$ _____

(c) Simplify $9(x + y)$ _____

(d) Simplify $3(c \cdot 7)$ _____

(e) $17 \times 0 \times 4 =$ _____

(f) What is the product of $2 \cdot a \cdot b \cdot 5$ _____

– – – – – – – – – – – – – – – –

(a) 7 and c
(b) 24
(c) $9x + 9y$
(d) $21c$
(e) 0
(f) $10ab$

If you did any of these incorrectly, make sure you find out where you went wrong. Correct your answer before you go on.

MULTIPLICATION AND SIGNS

③ Here are two rules that govern the multiplication of numbers with negative signs.

The multiplication of two quantities of like sign gives a result that is positive in sign.

Examples: $3 \times 3 = 9$
$(-3) \times (-3) = 9$

The multiplication of two quantities with differing signs gives a result that is negative in sign.

Examples: $3 \times (-3) = -9$
$(-1) \times 4 = -4$

Note: $(-1) \times 4 \times (-2) = +8$ because $(-1) \times 4 = -4$ and $(-4) \times (-2) = +8$.
And $(-1) \times (-4) \times (-2) = -8$ because $(-1) \times (-4) = +4$ and $(+4) \times (-2) = -8$.
Also $-5(a - 3) = -5a + 15$ because the factor outside the parentheses, -5, is multiplied to each term within the parentheses.
And $ab(4 - 4) = 0$ because $(4 - 4) = 0$ and thus is a zero factor.

PROBLEMS. Carry out the indicated operations.

(a) $3 \times 4 =$ _____

(b) $(-3)(-5) =$ _____

(c) $A \times B =$ _____

(d) $(-5)(a + b) =$ _____

(e) $(-a)(3 - b) =$ _____

(f) $W(a - a) =$ _____

— — — — — — — — — — — — — —

(a) $3 \times 4 = 12$
(b) $(-3)(-5) = +15$
(c) $A \times B = AB$
(d) $(-5)(a + b) = -5a - 5b$
(e) $(-a)(3 - b) = -3a + ab$
(f) $W(a - a) = W(0) = 0$

If you did any of these incorrectly, go back and see where you went wrong.

DIVISION TERMINOLOGY AND NOTATIONS

④ As with multiplication, most of us are familiar with the process of division, but need to review some basic terminology and the various ways of presenting the division operation. These are reviewed below.

divisor
   The quantity that is divided into another quantity.

Example: When 4 is divided into 8 to give 2, 4 is called the divisor.

dividend
The quantity that is divided by another quantity.

Example: When 4 is divided into 8 to give 2, 8 is the dividend.

quotient
The quantity that results from a division operation.

Example: When 4 is divided into 8 to give 2, 2 is called the quotient.

$\div$

Sign indicating that a division should be carried out. The quantity that precedes the sign is the dividend and the quantity that follows is the divisor. The sign means "divided by."

Example: $8 \div 4 = 2$ (read "8 divided by 4 equals 2")

$a\overline{)b}^{c}$

The sign $)$ indicates a division operation with the quantity outside of the sign, a, serving as the divisor. The quantity under the sign, b, is the dividend. The quotient, c, is written above the sign.

Example: $4\overline{)8}^{2}$ (4 is the divisor, 8 is the dividend, and 2 is the quotient)

$\dfrac{a}{b}$

Here the horizontal line indicates the division operation, with the quantity below the line serving as the divisor and the quantity above the line serving as the dividend.

Example: $\dfrac{8}{4} = 2$ (8 is the dividend, 4 is the divisor, and 2 is the quotient)

fraction
A division operation written in the general form $\dfrac{a}{b}$. The quantity on top of the line is called the numerator, and the quantity below the line is referred to as the denominator.

Example: $\dfrac{8}{4}$ is a fraction and is equal to 2; 8 is the numerator (dividend) and 4 is the denominator (divisor)

ratio
The generalized quantity, $\dfrac{a}{b}$, is often referred to as a ratio, read "the ratio of a to b"

Example: $\frac{8}{4}$ is the ratio of 8 to 4 and it equals 2.

decimal fraction
The result of carrying out a division operation when the denominator is larger than the numerator.

Examples: $\frac{2}{5}$ = 0.4 (0.4 is a decimal fraction)

$\frac{1}{4}$ = 0.25 (0.25 is a decimal fraction)

reciprocal
The reciprocal of a quantity is the quantity "turned over."

Examples: The reciprocal of $\frac{a}{b}$ is $\frac{b}{a}$.

The reciprocal of 2 is $\frac{1}{2}$ because $2 = \frac{2}{1}$ (a whole number is considered to have a denominator of 1).

The reciprocal of ab is $\frac{1}{ab}$.

inverse
The inverse of a quantity is the same as the reciprocal of that quantity.

per
Per means "in a" and it is a mathematical term for dividing. Thus, 60 miles per hour means you go 60 miles in one hour and is obtained by dividing miles by hours. For example, if we travel 120 miles in two hours, our speed is

$$\frac{120 \text{ miles}}{2 \text{ hours}} = 60 \text{ miles per hour}$$

Here we have divided the numerical factor in the denominator into itself and into the numerical factor in the numerator. The slash, /, is the abbreviation for "per", so 60 miles per hour might be abbreviated 60 miles/hour.

Examples: $\dfrac{10 \text{ miles}}{2 \text{ hours}} = \dfrac{\frac{10}{2} \text{ miles}}{\frac{2}{2} \text{ hours}} = \dfrac{5 \text{ miles}}{1 \text{ hour}} = 5 \text{ miles per hour}$

or 5 miles/hour

$\dfrac{48 \text{ bottles}}{2 \text{ cases}} = \dfrac{\frac{48}{2} \text{ bottles}}{\frac{2}{2} \text{ cases}} = \dfrac{24 \text{ bottles}}{1 \text{ case}} =$

24 bottles per case or 24 bottles/case

$$\frac{12 \text{ carbon atoms}}{3 \text{ molecules}} = \frac{\dfrac{12}{3} \text{ carbon atoms}}{\dfrac{3}{3} \text{ molecules}} =$$

$$\frac{4 \text{ carbon atoms}}{1 \text{ molecule}} = 4 \text{ carbon atoms per molecule or}$$

4 carbon atoms/molecule

$$\frac{1000 \text{ bacteria}}{4 \text{ plates}} = \frac{\dfrac{1000}{4} \text{ bacteria}}{\dfrac{4}{4} \text{ plates}} = \frac{250 \text{ bacteria}}{1 \text{ plate}}$$

= 250 bacteria per plate or 250 bacteria/plate

dividing out

A factor in the numerator may be divided by a factor in the denominator (other than zero), provided that neither of the factors is involved in a sum or difference and the factors are similar.

Example: $\dfrac{a}{a} = 1$ but $\dfrac{a}{b} = \dfrac{a}{b}$

This dividing out process may be used to simplify an expression as long as there are common factors in numerator and denominator. Often, "simplify" is the only instruction given for this process.

Examples: $\dfrac{3a}{a} = \dfrac{3\overset{1}{\cancel{a}}}{\underset{1}{\cancel{a}}} = 3 \times 1 = 3$ (The smaller, handwritten numbers show the results of the divding out.)

$\dfrac{12a}{4} = \dfrac{\overset{3}{\cancel{12}}a}{\underset{1}{\cancel{4}}} = 3a$

$\dfrac{4abR}{2bc} = \dfrac{\overset{2}{\cancel{4}}a\cancel{b}R}{\underset{1}{\cancel{2}\cancel{b}c}} = \dfrac{2aR}{c}$ (Often, when the result of dividing out is 1, the small number is not written, but just understood.)

$\dfrac{3R}{9Q} = \dfrac{\overset{1}{\cancel{3}}R}{\underset{3}{\cancel{9}}Q} = \dfrac{R}{3Q}$

$\dfrac{\cancel{R}(2R - Q)}{\cancel{R}} = 2R - Q$

Note: $\dfrac{1 + \cancel{a}b}{\cancel{a}} = \dfrac{1}{a} + \dfrac{\cancel{a}b}{\cancel{a}} = \dfrac{1}{a} + b.$

But $\dfrac{1 + \cancel{a}b}{\cancel{a}} \neq 1 + b$ ($\neq$ means "does not equal")

because $\dfrac{1 + ab}{a}$ means that *all* terms in $1 + ab$ must be divided by a.

Also $\dfrac{M(a + b)}{2M} = \dfrac{\cancel{M}(a + b)}{2\cancel{M}} = \dfrac{a + b}{2}$.

But $\dfrac{\cancel{4M}^2(a-b)}{\cancel{2M}_1 + 1} \neq 2(a-b)$. Here $2M$ cannot be used

in dividing out because it is part of a sum; the factor $(2M+1)$ could, however, be used in dividing out if the factor $(2M+1)$ appeared in the numerator.

$$\dfrac{2 \cancel{\text{yards}} \times 36 \text{ inches}}{1 \cancel{\text{yard}}} = 72 \text{ inches}$$

$$\dfrac{\cancel{3 \text{ moles}} \times 5 \text{ liters}}{\cancel{3 \text{ moles}}} = 5 \text{ liters}$$

$$\dfrac{\cancel{1 \text{ hour}} \times \cancel{120}^2 \text{ miles}}{\cancel{60} \text{ minutes} \times \cancel{1 \text{ hour}}} = 2 \text{ miles/minute}$$ (Hour was divid-

ed out but miles could not be divided out by minutes.)

PROBLEMS.

(a)   Perform the indicated operation $14 \div 7 =$ _____

(b)   Perform the indicated operation $4\overline{)12} =$ _____

(c)   If 3 is the divisor and 9 is the dividend, what is the quotient?   _____

(d)   The denominator of the fraction $\dfrac{6x}{2b}$ is   _____

(e)   The ratio of 7 to 11 would be written as   _____

(f)   The reciprocal of $\dfrac{4}{7}$ is   _____

(g)   The reciprocal of $4z$ is   _____

(h)   Express $\frac{3}{5}$ as a decimal fraction.   _____

(i)   Given $\dfrac{10 \text{ bleeps}}{2 \text{ minutes}}$, how many bleeps per minute?   _____

(j)   Simplify the expression $\dfrac{9xy(D+y)}{3y}$.   _____

(k) $\dfrac{6ab}{2a} =$ _____

(l) $\dfrac{3R(1+b)}{R} =$ _____

(m) $\dfrac{(a+1)(a-1)}{(a+1)} =$ _____

(n) $\dfrac{2+ab}{4b} =$ _____

- - - - - - - - - - - - - - - -

(a)   2
(b)   3
(c)   3

(d) $2b$

(e) $\dfrac{7}{11}$

(f) $\dfrac{7}{4}$

(g) $\dfrac{1}{4z}$

(h) $.6$

(i) $\dfrac{\cancel{10}^{5}\text{ bleeps}}{\cancel{2}\text{ minutes}} = \dfrac{5\text{ bleeps}}{1\text{ minute}} = 5\text{ bleeps per minute or 5 bleeps/minute}$

(j) $\dfrac{\cancel{3xy}^{3}(D+y)}{\cancel{3y}} = \dfrac{3x(D+y)}{1} = 3xD + 3xy$

(k) $\dfrac{\cancel{6ab}^{3}}{\cancel{2a}} = 3b$

(l) $\dfrac{3\cancel{R}(1+b)}{\cancel{R}} = 3(1+b) = 3 + 3b$

(m) $\dfrac{\cancel{(a+1)}(a-1)}{\cancel{a+1}} = a - 1$

(n) $\dfrac{2+ab}{4b} = \dfrac{\cancel{2}}{\cancel{4b}_{2}} + \dfrac{a\cancel{b}}{4\cancel{b}} = \dfrac{1}{2b} + \dfrac{a}{4}$

If you did any of these incorrectly, go back and see where you went wrong.

DIVISION AND SIGNS

⑤   Here are two rules that govern division by quantities involving negative signs.

Division of quantities with like signs gives a positive quotient.

Examples: $\dfrac{6}{3} = +2$

$\dfrac{-6}{-3} = +2$

$\dfrac{-b}{-a} = +\dfrac{b}{a}$

Division of quantities with unlike signs gives a negative quotient.

Examples: $\dfrac{-4}{2} = -2$

$\dfrac{4}{-2} = -2$

$$\frac{b}{-a} = -\frac{b}{a}$$

$$\frac{-b}{a} = -\frac{b}{a}$$

$\frac{a+b}{-c} = -\frac{a}{c} - \frac{b}{c}$ (because each term in the numerator is divided by the denominator)

$$\frac{-6xy}{3x} = -2y$$

$\frac{(-9)(x-y)}{3} = (-3)(x-y) = -3x + 3y$ (Note that dividing out was used here, and remember that a factor must be multiplied with each term within the parentheses.)

PROBLEMS. Carry out the indicated operations.

(a) $\frac{-3x}{-x} =$ _____

(b) $\frac{4(-3)}{2} =$ _____

(c) $\frac{(-4)(a-b)}{2} =$ _____

(d) $\frac{-5ab}{2b} =$ _____

- - - - - - - - - - - - - - - -

(a) 3

(b) −6

(c) −2a + 2b or −2(a − b)

(d) −2.5a or $\frac{5a}{2}$

If you would like to test yourself on the entire chapter, turn to the back of the book and take the final test for Chapter One. Otherwise, go on to the pretest and objectives for Chapter Two.

CHAPTER TWO

Math Operations on Fractions

.PRETEST AND OBJECTIVES

This pretest outlines the objectives for Chapter Two. It should help you identify what parts of the chapter, if any, you need to read. Try each of the problems which follow. Then check your answers with the answer key, and follow the directions given there.

Objectives and Pretest   **Your Answer**

When you complete this chapter, you will be able to:

1. Add or subtract simple fractions, finding a common denominator when necessary.

   (a) $\dfrac{1}{3} + \dfrac{2}{3} =$ _____

   (b) $\dfrac{1}{5} - \dfrac{2}{5} =$ _____

   (c) $\dfrac{3}{4} + \dfrac{2}{5} =$ _____

   (d) $\dfrac{1}{a} + \dfrac{2}{b} =$ _____

   (e) $\dfrac{3}{Q} - \dfrac{P}{4} =$ _____

   (f) $\dfrac{x}{2y} - \dfrac{3}{2y} =$ _____

   (g) $\dfrac{3}{a} + \dfrac{5}{2a} =$ _____

   (h) $\dfrac{J-2}{3} - \dfrac{X}{Y} =$ _____

(i) $\dfrac{3}{(a + b)} + \dfrac{4}{(a - b)}$

_____

2. Multiply simple fractions.

(a) $\dfrac{2}{6} \times \dfrac{2}{3} =$

_____

(b) $\dfrac{2}{7} \times \left(-\dfrac{1}{2}\right)\dfrac{P}{Q} =$

_____

(c) $(-5)\left(\dfrac{3a + b}{6c}\right) =$

_____

(d) $\dfrac{3(a + b)}{P} \times \dfrac{5P}{10(a + b)} =$

_____

(e) $\dfrac{7}{x + y} \times \dfrac{x - y}{14} =$

_____

(f) $\left(\dfrac{9}{x + y}\right)\left(\dfrac{A + C}{15}\right)\left(\dfrac{x + y}{6}\right) =$

_____

3. Divide simple and complex fractions.

(a) $\dfrac{1}{4} \div \dfrac{2}{5} =$

_____

(b) $\dfrac{P}{Q} \div \dfrac{C}{D} =$

_____

(c) $\dfrac{\dfrac{3}{4}}{\dfrac{1}{4}} =$

_____

(d) $\dfrac{\dfrac{a}{b}}{\dfrac{3}{c}} =$

_____

(e) $\dfrac{\dfrac{5}{6}}{\dfrac{\dfrac{2}{3}}{\dfrac{1}{a}}} =$

_____

(f) $\dfrac{R}{\dfrac{1}{9}} =$

_____

(g) $\dfrac{\dfrac{1}{3} + \dfrac{7}{6} + \dfrac{3}{4}}{\dfrac{3}{2}} =$

_____

(h) $\dfrac{\dfrac{a+b}{3}}{\dfrac{3(a+b)}{2}} = $ _____

(i) $\dfrac{\dfrac{pounds}{ounces}}{\dfrac{pounds}{dollar}} = $ _____

Answer Key and Directions

Check your answers with the answer key which follows. If you missed any of these questions or if you want help with a particular objective, read the indicated portion of Chapter Two.

If you missed any of these problems: See frames:

1. (a) $\dfrac{3}{3}$ or 1

(b) $-\dfrac{1}{5}$

(c) $\dfrac{23}{20}$ or $1\frac{3}{20}$

(d) $\dfrac{b+2a}{ab}$

(e) $\dfrac{12-PQ}{4Q}$

(f) $\dfrac{x-3}{2y}$

(g) $\dfrac{11}{2a}$

(h) $\dfrac{JY-2Y-3x}{3Y}$ or $\dfrac{Y(J-2)-3X}{3Y}$

(i) $\dfrac{7a+b}{(a+b)(a-b)}$ frame 1, page 18

2. (a) $\dfrac{4}{18}$ or $\dfrac{2}{9}$

(b) $-\dfrac{P}{7Q}$

(c) $\dfrac{-15a-5b}{6c}$ or $-\dfrac{(15a+5b)}{6c}$

(d) $\dfrac{3(a+b)5P}{10P(a+b)} = \dfrac{\overset{3}{\cancel{15}}}{\underset{2}{\cancel{10}}} = \dfrac{3}{2}$

(e) $\dfrac{\cancel{7}(x-y)}{(x+y)\cancel{14}_2} = \dfrac{x-y}{2(x+y)}$

(f) $\dfrac{\cancel{9}(A+C)\cancel{(x+y)}}{\cancel{(x+y)}\,(\cancel{15})\,\cancel{6}} = \dfrac{A+C}{10}$

frame 2, page 24

3. (a) $\dfrac{5}{8}$

(b) $\dfrac{DP}{CQ}$

(c) 3

(d) $\dfrac{ac}{3b}$

(e) $\dfrac{15}{12a}$ or $\dfrac{5}{4a}$

(f) $9R$

(g) $1\frac{1}{2}$

(h) $\dfrac{2}{9}$

(i) $\dfrac{\text{dollar}}{\text{ounces}}$

frame 3, page 25

ADDITION AND SUBTRACTION

1 To add or subtract fractions, the fractions must have the same quantity in the denominator. When two fractions have the same quantity in the denominator, they are said to have a *common denominator*.

When adding or subtracting fractions with a common denominator, add or subtract the quantities in the numerators and leave the value of the denominator unchanged.

Examples:

$$\frac{1}{4} + \frac{2}{4} = \frac{1+2}{4} = \frac{3}{4}$$

$$\frac{2}{3} - \frac{1}{3} = \frac{2-1}{3} = \frac{1}{3}$$

$$\frac{1}{5} - \frac{3}{5} = -\frac{2}{5}$$

$$\frac{a}{b} + \frac{c}{b} = \frac{a+c}{b}$$

$$\frac{4}{5x} - \frac{y+z}{5x} = \frac{4-(y+z)}{5x} \text{ or } \frac{4-y-z}{5x}$$

If two fractions have different quantities in the denominator, we must give them a common denominator before we can add or subtract them. The following procedure is used to find a common denominator for these fractions which are being added.

$$\frac{1}{3} + \frac{3}{4}$$

(1)   Write the fractions separately.

$$\frac{1}{3}$$

$$\frac{3}{4}$$

(2)   Multiply the denominators of the two initial fractions to give a common denominator for the two new fractions. Thus the new denominator is the product of the two initial denominators.

$$\frac{1}{3} = \frac{}{12}$$

$$(3 \times 4 = 12)$$

$$\frac{3}{4} = \frac{}{12}$$

(3)   Divide, in turn, the initial denominators into the new denominator and multiply the result with the initial numerator.

$$\frac{1}{3} = \frac{4 \times 1}{12} = \frac{4}{12}$$

3 (the initial denominator) divided into 12 (the new denominator) gives 4, which is multiplied by the initial numerator, 1, giving $\frac{4}{12}$.

$$\frac{3}{4} = \frac{3 \times 3}{12} = \frac{9}{12}$$

4 (the initial denominator) divided into 12 (the new denominator) gives 3, which is multiplied by the initial numerator, 3, giving $\frac{9}{12}$.

(4)   Now we have new fractions that have a common denominator and that can be added or subtracted.

$$\frac{4}{12} + \frac{9}{12} = \frac{13}{12}$$

Examples: $\frac{1}{2} + \frac{2}{3}$

$$\frac{1}{2} = \frac{3 \times 1}{6} = \frac{3}{6}$$

$$\frac{2}{3} = \frac{2 \times 2}{6} = \frac{4}{6}$$

$$\frac{1}{2} + \frac{2}{3} = \frac{3}{6} + \frac{4}{6} = \frac{7}{6}$$

Example: $\frac{3}{a} + \frac{2}{b}$

$$\frac{3}{a} = \frac{3b}{ab}$$

$$\frac{2}{b} = \frac{2a}{ab}$$

$$\frac{3}{a} + \frac{2}{b} = \frac{3b}{ab} + \frac{2a}{ab} = \frac{3b + 2a}{ab}$$

Example: $\frac{6}{P} - \frac{W}{Q}$

$$\frac{6}{P} = \frac{6Q}{PQ}$$

$$\frac{W}{Q} = \frac{PW}{PQ}$$

$$\frac{6}{P} - \frac{W}{Q} = \frac{6Q}{PQ} - \frac{PW}{PQ} = \frac{6Q - PW}{PQ}$$

If you can see that one of the initial denominators is a multiple of the other, there is no need to find a new denominator because the common denominator is already defined. In that case, multiply both numerator and denominator of the fraction by the appropriate multiple, so that the fractions have the same denominator.

Example: $\frac{1}{4} - \frac{1}{8}$

$$\frac{1}{4} = \frac{2 \times 1}{8} = \frac{2}{8}$$

$$\frac{1}{8} \text{ remains unchanged}$$

$$\frac{1}{4} - \frac{1}{8} = \frac{2}{8} - \frac{1}{8} = \frac{2 - 1}{8} = \frac{1}{8}$$

Example: $\frac{2}{c} + \frac{3}{4c}$

$$\frac{2}{c} = \frac{2 \times 4}{4c} = \frac{8}{4c}$$

$$\frac{3}{4c} \text{ remains unchanged}$$

$$\frac{2}{c} + \frac{3}{4c} = \frac{8}{4c} + \frac{3}{4c} = \frac{11}{4c}$$

(5) Often, your answer will not be simplified. To simplify, you should multiply out (distribute any multiplication) the numerator obtained and combine similar quantities.

Examples: $\dfrac{M-3}{6} - \dfrac{K}{L}$

$$\frac{M-3}{6} = \frac{L(M-3)}{6L}$$

$$\frac{K}{L} = \frac{6K}{6L}$$

$$\frac{M-3}{6} - \frac{K}{L} = \frac{L(M-3)}{6L} - \frac{6K}{6L} = \frac{L(M-3)-6K}{6L} =$$

$$\frac{LM-3L-6K}{6L}$$

(Note: After we multiplied out the numerator, none of the quantities could be combined because they weren't similar.)

$$\frac{2}{x+y} - \frac{5}{x-y}$$

$$\frac{2}{x+y} = \frac{2(x-y)}{(x+y)(x-y)}$$

$$\frac{5}{x-y} = \frac{5(x+y)}{(x-y)(x+y)}$$

$$\frac{2}{x+y} - \frac{5}{x-y} = \frac{2(x-y)}{(x+y)(x-y)} - \frac{5(x+y)}{(x-y)(x+y)} =$$

$$\frac{2(x-y)-5(x+y)}{(x+y)(x-y)} = \frac{2x-2y-5x-5y}{(x+y)(x-y)} =$$

$$\frac{-3x-7y}{(x+y)(x-y)}$$

(6) Sometimes, the answer can be divided out because the common denominator was not the lowest common denominator. Therefore, always check to see if you can divide out (simplify) any fraction answer.

Example: $\dfrac{1}{6} + \dfrac{1}{8}$

$$\frac{1}{6} = \frac{8 \times 1}{6 \times 8} = \frac{8}{48}$$

$$\frac{1}{8} = \frac{6 \times 1}{6 \times 8} = \frac{6}{48}$$

$$\frac{8}{48} + \frac{6}{48} = \frac{14}{48} = \frac{14/\!\!/2}{48/\!\!/2} = \frac{7}{24}$$

(7) Finally, when adding or subtracting mixed numbers (those with a whole number and a fraction), always convert to improper fractions

before proceeding. To convert a mixed number to an improper fraction, multiply the denominator by the number and then add this result to the numerator. Your result is the new numerator.

Example: $3\frac{2}{11} + \frac{1}{2} = \frac{35}{11} + \frac{1}{2}$ (Note: $3\frac{2}{11}$ is $\frac{35}{11}$ because

$$11 \times 3 = 33 \text{ and } 33 + 2 = 35.)$$

$$\frac{35}{11} = \frac{35 \times 2}{22} = \frac{70}{22}$$

$$\frac{1}{2} = \frac{11}{22}$$

$$3\frac{2}{11} + \frac{1}{2} = \frac{35}{11} + \frac{1}{2} = \frac{70}{22} + \frac{11}{22} = \frac{81}{22}$$

(Note: Answers can be left as improper fractions.)

PROBLEMS. Carry out the indicated operations.

(a) $\frac{2}{3} + \frac{4}{3} =$ _____

(b) $\frac{1}{7} - \frac{3}{7} =$ _____

(c) $\frac{2}{5} + \frac{1}{3} =$ _____

(d) $\frac{4}{x} + \frac{3}{y} =$ _____

(e) $\frac{2}{S} - \frac{T}{3} =$ _____

(f) $\frac{3}{a} + \frac{1}{3a} =$ _____

(g) $\frac{1}{xy} + \frac{2}{xy} =$ _____

(h) $\frac{L-2}{4} - \frac{R}{Q} =$ _____

(i) $\frac{4}{a+b} - \frac{3}{a-b} =$ _____

- - - - - - - - - - - - - -

(a) $\frac{2+4}{3} = \frac{6}{3} = 2$

(b) $\frac{1-3}{7} = -\frac{2}{7}$

(c)
$$\frac{2}{5} = \frac{2 \times 3}{15} = \frac{6}{15}$$

$$\frac{1}{3} = \frac{1 \times 5}{15} = \frac{5}{15}$$

$$\frac{2}{5} + \frac{1}{3} = \frac{6}{15} + \frac{5}{15} = \frac{11}{15}$$

(d)
$$\frac{4}{x} = \frac{4y}{xy}$$

$$\frac{3}{y} = \frac{3x}{xy}$$

$$\frac{4}{x} + \frac{3}{y} = \frac{4y}{xy} + \frac{3x}{xy} = \frac{4y + 3x}{xy}$$

(e)
$$\frac{2}{S} = \frac{2 \times 3}{3S} = \frac{6}{3S}$$

$$\frac{T}{3} = \frac{ST}{3S}$$

$$\frac{2}{S} - \frac{T}{3} = \frac{6}{3S} - \frac{ST}{3S} = \frac{6 - ST}{3S}$$

(f)
$$\frac{3}{a} = \frac{3 \times 3}{3a} = \frac{9}{3a}$$

$$\frac{1}{3a} = \frac{1 \times 1}{3a} = \frac{1}{3a}$$

$$\frac{3}{a} + \frac{1}{3a} = \frac{9}{3a} + \frac{1}{3a} = \frac{9 + 1}{3a} = \frac{10}{3a}$$

(g)
$$\frac{1 + 2}{xy} = \frac{3}{xy}$$

(h)
$$\frac{L - 2}{4} = \frac{Q(L - 2)}{4Q}$$

$$\frac{R}{Q} = \frac{4R}{4Q}$$

$$\frac{L - 2}{4} - \frac{R}{Q} = \frac{Q(L - 2)}{4Q} - \frac{4R}{4Q} = \frac{Q(L - 2) - 4R}{4Q} = \frac{QL - 2Q - 4R}{4Q}$$

(i)
$$\frac{4}{a + b} = \frac{4(a - b)}{(a + b)(a - b)}$$

$$\frac{3}{a - b} = \frac{3(a + b)}{(a + b)(a - b)}$$

$$\frac{4}{a + b} - \frac{3}{a - b} = \frac{4(a - b)}{(a + b)(a - b)} - \frac{3(a + b)}{(a + b)(a - b)} =$$

$$\frac{4(a - b) - 3(a + b)}{(a + b)(a - b)} = \frac{4a - 4b - 3a - 3b}{(a + b)(a - b)} = \frac{a - 7b}{(a + b)(a - b)}$$

MULTIPLICATION OF FRACTIONS

2 To multiply two fractions, multiply the numerators by the numerators, and the denominators by the denominators. Simplify your answer by dividing out or combining similar quantities after multiplying out.

Examples:
$$\frac{1}{2} \times \frac{2}{3} = \frac{1 \times 2}{2 \times 3} = \frac{2}{6} \text{ or } \frac{1}{3}$$

$$\left(\frac{4}{5}\right)\left(-\frac{1}{2}\right)\left(\frac{a}{b}\right) = -\frac{4 \times 1 \times a}{5 \times 2 \times b} = -\frac{4a}{10b} \text{ or } -\frac{2a}{5b}$$

$$4\left(-\frac{a+1}{b}\right) = \frac{4}{1}\left(-\frac{a+1}{b}\right) = -\frac{4(a+1)}{1 \times b} = -\frac{4a+4}{b}$$

$$5\left(\frac{2}{3}\right) = \frac{10}{3}$$

$$\frac{4a}{9b} \cdot \frac{1x}{2y} = \frac{4ax}{18by} \text{ or } \frac{2ax}{9by}$$

$$\frac{s+t}{6} \cdot \frac{1}{5(s+t)} = \frac{(s+t)}{6(5)(s+t)} = \frac{1}{30}$$

PROBLEMS. Carry out the following multiplications.

(a) $\dfrac{2}{7} \times \dfrac{1}{3} =$ _____

(b) $\dfrac{4a}{7} \times \dfrac{2b}{3} =$ _____

(c) $8\left(-\dfrac{4}{5}\right) =$ _____

(d) $(-7)\left(\dfrac{a}{b}\right)\left(-\dfrac{2}{3}\right) =$ _____

(e) $\dfrac{b}{a}\left(\dfrac{4+1}{c}\right) =$ _____

(f) $\dfrac{11(2x+3y)}{12} \cdot \dfrac{3(a+b)}{22} =$ _____

- - - - - - - - - - - - - - - -

(a) $\dfrac{2}{21}$

(b) $\dfrac{8ab}{21}$

(c) $-\dfrac{32}{5}$

(d) $\dfrac{14a}{3b}$

(e) $\dfrac{4b + b}{ac} = \dfrac{5b}{ac}$

(f) $\dfrac{\cancel{33}(2x + 3y)(a + b)}{\cancel{(12)}\ \cancel{(22)}}\ = \dfrac{(2x + 3y)(a + b)}{8}$

DIVISION OF FRACTIONS

③   To divide fractions, we "invert and multiply." That means when the operation is written out with the division sign, the reciprocal of the divisor is multiplied by the dividend. (Remember, the reciprocal is obtained by inverting—"turning over"—a number.)

Examples: $\dfrac{1}{3} \div \dfrac{1}{6} = \dfrac{1}{3} \times \dfrac{6}{1} = \dfrac{6}{3} = 2$

$\dfrac{3}{4} \div \dfrac{2}{5} = \dfrac{3}{4} \times \dfrac{5}{2} = \dfrac{15}{8}$ or $1\dfrac{7}{8}$

$\dfrac{x}{y} \div \dfrac{a}{b} = \dfrac{x}{y} \times \dfrac{b}{a} = \dfrac{bx}{ay}$

When the division operation is written as a fraction, we take the reciprocal of the fraction in the denominator and multiply it by the fraction in the numerator.

Examples: $\dfrac{\frac{1}{2}}{\frac{1}{4}} = \dfrac{4}{1} \times \dfrac{1}{2} = \dfrac{4}{2} = 2$

$\dfrac{\frac{4}{5}}{\frac{2}{10}} = \dfrac{10}{2} \times \dfrac{4}{5} = \dfrac{40}{10} = 4$

$\dfrac{\frac{b}{a}}{\frac{b}{c}} = \dfrac{c}{b} \times \dfrac{b}{a} = \dfrac{cb}{ba} = \dfrac{c}{a}$

$\dfrac{\frac{2}{4}}{\frac{\frac{1}{2}}{\frac{1}{8}}} = \dfrac{\frac{2}{4}}{\frac{8}{1} \times \frac{1}{2}} = \dfrac{\frac{2}{4}}{\frac{8}{2}} = \dfrac{2}{8} \times \dfrac{2}{4} = \dfrac{4}{32} = \dfrac{1}{8}$

(Note: We did the division in the denominator first because the long bar placed two fractions in the denominator. It acts like parentheses and operations inside parentheses are always done first.)

$$\frac{\dfrac{a}{b}}{\dfrac{c}{d}} = \frac{\dfrac{a}{b}}{\dfrac{d}{f} \times \dfrac{c}{d}} = \frac{\dfrac{a}{b}}{\dfrac{dc}{fd}} = \frac{\dfrac{a}{b}}{\dfrac{c}{f}} = \frac{f}{c} \times \frac{a}{b} = \frac{af}{bc}$$

$$\frac{\dfrac{1}{2} + \dfrac{3}{4} + \dfrac{5}{8}}{\dfrac{2}{3}} = \frac{3}{2}\left(\frac{1}{2} + \frac{3}{4} + \frac{5}{8}\right) = \frac{3 \times 1}{2 \times 2} + \frac{3 \times 3}{2 \times 4} + \frac{3 \times 5}{2 \times 8}$$

$$= \frac{3}{4} + \frac{9}{8} + \frac{15}{16} = \frac{12}{16} + \frac{18}{16} + \frac{15}{16} = \frac{45}{16}$$

$$\frac{\dfrac{x+1}{2}}{\dfrac{x-1}{3}} = \frac{x+1}{2} \times \frac{3}{x-1} = \frac{3(x+1)}{2(x-1)}$$

$$\frac{\dfrac{\text{miles}}{\text{minutes}}}{\dfrac{\text{miles}}{\text{kilometers}}} = \frac{\text{miles}}{\text{minutes}} \times \frac{\text{kilometers}}{\text{miles}} = \frac{\text{kilometers}}{\text{minutes}}$$

(Note: Divide out factors where possible.)

$$\frac{\dfrac{a+2}{2}}{\dfrac{4(a+2)}{3}} = \frac{(a+2)}{2} \times \frac{3}{4(a+2)} = \frac{3}{8}$$

The process of dividing fractions is called *simplifying complex fractions*.

A similar procedure is used to simplify complex fractions such as:

$$\frac{5}{\dfrac{1}{5}} \, .$$

The trick here, is to realize that:

$$\frac{5}{\dfrac{1}{5}} = \frac{\dfrac{5}{1}}{\dfrac{1}{5}}$$

and to solve as before.

$$\frac{\dfrac{5}{1}}{\dfrac{1}{5}} = \frac{5}{1} \times \frac{5}{1} = 25$$

PROBLEMS. Carry out the indicated operations.

(a) $\dfrac{5}{2} \div \dfrac{3}{4} =$ _____

(b) $\dfrac{x}{y} \div \dfrac{3a}{2} =$ _____

(c) $\dfrac{\frac{2}{3}}{\frac{1}{9}} =$ _____

(d) $\dfrac{\frac{\frac{5}{8}}{\frac{1}{2}}}{\frac{3a}{4}} =$ _____

(e) $\dfrac{\frac{1}{2} + \frac{3}{4} + \frac{4}{3}}{\frac{2}{3}} =$ _____

(f) $\dfrac{\frac{a}{b}}{\frac{a+1}{4}} =$ _____

(g) $\dfrac{M}{\frac{1}{3}} =$ _____

(h) $\dfrac{\frac{grams}{liters}}{\frac{grams}{mole}} =$ _____

(i) $\dfrac{\frac{10a}{b}}{\frac{2}{b}} =$ _____

— — — — — — — — — — — — — — —

(a) $\dfrac{10}{3}$

(b) $\dfrac{2x}{3ay}$

(c) 6

(d) $\dfrac{\dfrac{\cancel{2}}{1} \times \dfrac{5}{\cancel{8}\,4}}{\dfrac{3a}{4}} = \dfrac{\dfrac{5}{4}}{\dfrac{3a}{4}} = \dfrac{\cancel{4}}{3a} \times \dfrac{5}{\cancel{4}} = \dfrac{5}{3a}$

(Note: Since the long bar left two fractions in the numerator, they must be divided first.)

(e) $\dfrac{3}{2}\left(\dfrac{1}{2} + \dfrac{3}{4} + \dfrac{4}{3}\right) = \dfrac{3 \times 1}{2 \times 2} + \dfrac{3 \times 3}{2 \times 4} + \dfrac{3 \times 4}{2 \times 3} = \dfrac{3}{4} + \dfrac{9}{8} + \dfrac{12}{6} = \dfrac{3}{4} + \dfrac{9}{8} + 2$

$= \dfrac{6}{8} + \dfrac{9}{8} + 2 = \dfrac{15}{8} + 2 = 1\tfrac{7}{8} + 2 = 3\tfrac{7}{8}$

(f) $\left(\dfrac{4}{a+1}\right)\left(\dfrac{a}{b}\right) = \dfrac{4a}{ab+b}$

(g) $\dfrac{\dfrac{M}{1}}{\dfrac{1}{3}} = \dfrac{3}{1} \times \dfrac{M}{1} = 3M$

(h) $\left(\dfrac{\text{mole}}{\cancel{\text{grams}}}\right)\left(\dfrac{\cancel{\text{grams}}}{\text{liter}}\right) = \dfrac{\text{mole}}{\text{liter}}$

(i) $\dfrac{\cancel{b}}{\cancel{2}} \times \dfrac{\overset{5}{\cancel{10a}}}{\cancel{b}} = 5a$

CHAPTER THREE

Algebra

PRETEST AND OBJECTIVES

This pretest outlines the objectives for Chapter Three. It should help you identify what parts of the chapter, if any, you need to read. Try each of the problems which follow. Then check your answers with the answer key, and follow the directions given there.

Objectives and Pretest   **Your Answer**

When you complete this chapter, you will be able to:

1. Apply and interpret these signs: $=, >, <$.
   For problems (a) through (d) fill in
   either $=, >,$ or $<$.

   (a) $5a$ _____ $3a + 2a$

   (b) $\dfrac{12}{-4}$ _____ $\dfrac{-15}{-3}$

   (c) $-2B$ _____ B if $B < 0$

   (d) X _____ $\dfrac{X}{B}$ if $X > 0, B > 1$

2. Write an algebraic equation from a written
   sentence. Do not worry about the symbols
   you use. Write an algebraic equation for
   each of the following.

   (a) Growth is equal to birth rate minus death
        rate.   _____

   (b) Speed is the ratio of distance to time.   _____

   (c) The fraction of sulfur atoms in hydrogen
        sulfide is the ratio of sulfur atoms to the
        total number of hydrogen and sulfur atoms.   _____

3. Substitute values for variables in an equation and solve for the indicated variable.

    (a)   Identify the variables in the equation $M = 3X + 2$

    (b)   Given the above equation and that $X = 4$, what is M?

4. Identify the independent variable (the given) and the dependent variable (the unknown) in a given equation.

   In the equation $A = 3B - 1$:

    (a)   Which is the independent variable?

    (b)   Which is the dependent variable?

5. Solve equations for the indicated variable, rearranging terms when necessary.

    (a) $4X = Y$, solve for X

    (b) $R = \dfrac{2P}{Q}$, solve for Q, where $Q \neq 0$ and $R \neq 0$

    (c) $5X - 2 = Y$, solve for X

    (d)   solve $B = \dfrac{5}{C}(-4) + 20$ for C, where $C \neq 0$ and $B \neq 20$

    (e)   solve $6(X - 4) = Y$ for X

    (f)   solve $^\circ C = \dfrac{5}{9}(^\circ F - 32^\circ)$ for $^\circ F$

    (g) $4a - 4b = c$, solve for a

    (h) $2t - Qt = y + 4$, solve for t

    (i) $H(M_1 - M_2) = R(M_1 - M_3)$, solve for M_1

6. Write an algebraic equation from a written sentence, substitute values for the variables in the equation, and solve for the indicated variable, rearranging when necessary.

    (a)   In a series of compounds composed of carbon, hydrogen, and nitrogen, it is found that the number of hydrogen atoms is defined by twice the number of carbon atoms plus three. If a molecule in this series has twenty-one atoms of hydrogen, how many carbon atoms will it have?

_____

(b) A child's IQ score is 100 times the ratio of his mental age to his chronological age. If the child is 5 years old chronologically and his IQ score is 120, what is his mental age?

_____

Answer Key and Directions

Check your answers with the answer key which follows. If you had difficulty working any of the problems or if you want help with a particular objective, read the indicated portion of Chapter Three.

If you missed any of these problems: **See frames**

1. (a)   =
   (b)   <
   (c)   >
   (d)   >   frame 1, page 32

2. Your answers are correct if your symbols are different from those given below, as long as the form of your equations is the same.

   (a) $G = B - D$, where G = growth,
         B = birth rate, and D = death rate

   (b) $s = \dfrac{d}{t}$, where s = speed, d = distance
         and t = time.

   (c) $F_s = \dfrac{S}{S + H}$, where F_s = the fraction of
         sulfur atoms, S = the number of sulfur
         atoms, and H = the number of hydrogen
         atoms.   frame 2, page 34

3. (a) M and X are variables
   (b)   14   frames 3-4, page 35

4. (a) B
   (b) A frame 5, page 36

5. (a) $X = \dfrac{Y}{4}$

   (b) $Q = \dfrac{2P}{R}$

   (c) $X = \dfrac{Y + 2}{5}$ or $\dfrac{Y}{5} + \dfrac{2}{5}$

(d) $C = -\dfrac{20}{B - 20}$

(e) $X = \dfrac{Y}{6} + 4$

(f) $°F = \dfrac{9}{5} °C + 32°$

(g) $a = \dfrac{c + 4b}{4}$

(h) $t = \dfrac{y + 4}{2 - Q}$

(i) $M_1 = \dfrac{HM_2 - RM_3}{H - R}$

frames 6-14, page 38

6. (a) $H = 2C + 3$, where H = hydrogen atoms and C = carbon atoms.
$H = 21$, so
$21 = 2C + 3$
$18 = 2C$
$9 = C$

(b) $IQ = 100\left(\dfrac{MA}{CA}\right)$, where MA = mental age and CA = chronological age.
$IQ = 120$ and
$CA = 5$, so
$120 = \overset{20}{\cancel{100}}\left(\dfrac{MA}{\cancel{5}}\right)$
$120 = 20\,MA$
$6 = MA$

frames 15-17, page 46

THE EQUATION IDEA

① One of the most important mathematical skills required in the sciences is the ability to work problems in the abstract, that is, using symbols instead of numbers. To accomplish this, one must be familiar with the basic operations and principles of algebra, which we will review in this chapter.

The first principle of algebra is the equation. An equation consists of three parts: the left hand side, the equals sign, and the right hand side. Left = Right. This means that whatever is written on the left side of the equation is the same as the right side even though it may not look the same. Consider the following equation.

$1 + 1 = 2$

The left side does not look like the right side, but they are equivalent since they are both equal to 2.

An untrue equation—or an inequality—would be:

$$1 + 1 = 3$$

and we would say $1 + 1 \neq 3$, where the slash mark means an unequal relationship. Another notation that is often used in this case is $1 + 1 < 3$, where $<$ means the total of all quantities on the left is *less than* the total of all quantities on the right. If we had written $3 > 1 + 1$, the $>$ means *greater than*, that is 3 is greater than $1 + 1$.

Unlike other division, division by proper fractions (those between 0 and 1) results in larger magnitude.

Examples: $7 < \dfrac{7}{\frac{1}{3}}$ since $\dfrac{7}{\frac{1}{3}} = (7)(3) = 21$

$x < \dfrac{x}{y}$.   where $x > 0$ and y is a proper fraction $(0 < y < 1)$.

The larger the magnitude of a negative number, the smaller it is.

Examples: $-4 < -1$, even though $4 > 1$.
$12\,x < x$ when $x < 0$ because $12(-1) < -1, 12(-2) < 2,....$

PROBLEMS. Indicate which sign (=, >, or <) correctly completes the following statements.

(a)   5 _____ 3 + 2

(b)   15 _____ 3 x 5

(c)   10 _____ 2 x 6

(d) $\dfrac{10}{-2}$ _____ $\dfrac{-14}{-2}$

(e) $A + 2$ _____ A

(f) $(-2)(-2)$ _____ $(-2)(-2)(-2)$

(g) $-5A$ _____ A, if $A > 0$

(h) X _____ $\dfrac{X}{Y}$, if $X > 0$ and $Y > 1$

(i) X _____ $\dfrac{X}{Y}$, if $X < 0$ and $0 < y < 1$

— — — — — — — — — — — — — —

(a) $5 = 3 + 2$
(b) $15 = 3 \times 5$
(c) $10 < 2 \times 6, (2 \times 6 = 12)$
(d) $\dfrac{10}{-2} < \dfrac{-14}{-2}$. Remember, when we divide numbers of unlike signs the result is always negative, and division of numbers with like signs is positive. This equation is the same as $-5 < +7$.
(e) $A + 2 > A$
(f) $(-2)(-2) > (-2)(-2)(-2)$, because $(-2)(-2) = +4$ and $(-2)(-2)(-2) = -8$, and $4 > -8$.
(g) $-5A < A$. If $A = 0$ then $(-5)(0) = 0$ and if A is less than zero, say -1, then $(-5)(-1) > -1$ because $(-5)(-1) = +5$.
(h) $X > \dfrac{X}{Y}$

(i) $x < \dfrac{x}{y}$, because although $\dfrac{x}{y}$ has a larger magnitude than x, both x and $\dfrac{x}{y}$ are negative, which means $\dfrac{x}{y}$ is smaller.

If you did any of these incorrectly, make sure you find out where you went wrong.

SYMBOLS AND ALGEBRAIC STATEMENTS

(2)   To convert a sentence into an algebraic equation, you must represent the words that are to be variables with letters. Usually, you would use the first letter in the word but you can establish your own rules.

> Example: Volume of a box equals the length times the width times the height.
> v = volume
> l = length
> w = width
> h = height

Also, you must be aware of mathematical words that give you the key relationships within the equation. Examples of such words are *add, subtract, multiply, divide, sum, difference, product, quotient, times, ratio* and *equals*.

> Examples: The death rate in a fruit fly jar is the number of dead flies divided by the number of days.
> D = death rate
> F = dead flies
> N = number of days
> $D = \dfrac{F}{N}$
>
> FDR was president 4 times longer than President Kennedy
> R = FDR
> K = Kennedy
> $R = 4K$
>
> Force equals mass times acceleration.
> f = force
> m = mass
> a = acceleration
> $f = ma$

PROBLEMS. Express the following statements as algebraic equations.

(a)   Distance is the product of speed and time. _____

(b)   Degrees Kelvin is equal to the sum of
degrees Centigrade and 273 degrees. _____

(c) The fraction of psychotic individuals in
the general population is equal to the ratio
of psychotics to the sum of psychotic and
normal individuals.   _____

(d) The total number of carbon atoms in an
alkane molecule is less than the total
number of hydrogen atoms.   _____

– – – – – – – – – – – – – – – –

Do not worry if your symbols and those in the answers are not identi-
cal. Just make sure that they have the same meaning.

(a) $D = ST$, where D = distance, S = speed, and T = time
(b) °K = °C + 273°C, where °K = degrees Kelvin and °C = degrees
Centigrade
(c) $F_P = \dfrac{P}{P + N}$, where F_P = fraction of psychotics, P = number of
psychotics, and N = the number of normal individuals
(d) $C_T < H_T$ where C_T = total number of carbon atoms and H_T = total
number of hydrogen atoms.

SUBSTITUTION OF VALUES FOR VARIABLES

(3)   An important property of equations is that they allow us to relate
varying quantities by memorizing one general statement instead of a
table of relationships. For example, to convert temperature in degrees
Fahrenheit to degrees Centigrade, we can do one of two things: construct a table relating the two temperature scales, or memorize the
equation below.

$$°C = \frac{5}{9} \, (°F - 32°)$$

In this equation we see that the symbol °C stands for degrees Centigrade, and the symbol °F stands for degree Fahrenheit. Since the terms
°C and °F can change, they are called *variables*. The numbers $\frac{5}{9}$ and
−32° do not change and are called *constants*.
To use this equation, as it is written, just substitute a value for the
variable, °F, and carry out the indicated calculation to find the value
for the °C variable.

Example: To convert 98.6°F (human body temperature) into
degrees Centigrade, we substitute 98.6°F for °F in the
equation and complete the indicated calculation.

$$°C = \frac{5}{9} \, (°F - 32°)$$

$$°C = \frac{5}{9} \, (98.6° - 32°)$$

$$°C = \frac{5}{9}\ (66.6°)$$

$$°C = \frac{333}{9}$$

$$°C = 37°$$

PROBLEM. Given the equation below:

$$D = \frac{M}{V}$$

where D = density, M = mass, and V = volume, calculate the density of a substance that has a mass of 16 grams and occupies a volume of 0.04 liters. Use the space below or a sheet of paper for your calculations.

—————————————

M = 16 grams, V = 0.04 liters

$$D = \frac{16 \text{ grams}}{0.04 \text{ liters}}$$

$$D = \frac{400 \text{ grams}}{\text{liter}} \text{ or } D = 400 \text{ grams/liter}$$

Notice that in this problem, all three terms — D, M, and V — are variables.

④ PROBLEM. The perimeter of a rectangle equals the sum of twice the width and twice the length. If the perimeter is 33 and the length is $9\frac{1}{2}$, what is the width? Use the space below to establish your variables and equation — and do your calculations.

—————————————

p = perimeter, l = length, w = width

$p = 2w + 2l$

$p = 33,\ l = 9\frac{1}{2}$, so

$33 = 2w + 19$

$14 = 2w$

$7 = w$

SOLVING EQUATIONS: THE DEPENDENT VARIABLE

⑤ In frame 3 we said that the equation:

$$°C = \frac{5}{9}\ (°F - 32°),$$

has two variables — the terms °C and °F. To be more specific we should say that °C is a *dependent variable* because, as the equation is written, the value of °C is dependent on a value of °F which is the *independent variable*.

Most problems that lend themselves to algebraic solutions will furnish *given* information. This given information corresponds to the independent variables. The *unknown*, or what is being solved for, is the dependent variable. For this reason, we put the unknown or dependent variable on one side of the equation and all other terms (independent variables and constants) on the other side.

Examples: In the equation $A = 3X - \frac{7}{8}$,

The independent variable is X.
The dependent variable is A.

(Remember, 3 and $\frac{7}{8}$ are constants.)

In the equation $z = 2x + 3y + 15$,
x and y are independent variables.
z is the dependent variable.
2, 3, and 15 are constants.

All of these concepts are rather straightforward, but what if you are given the dependent variable and are asked to find the value of the independent variable? For example, if ice melts at 0°C, at what temperature does ice melt on the Fahrenheit scale?

To answer this question, you can do one of two things: memorize a new equation, $°F = \frac{9}{5} °C + 32°$, where the dependent variable is °F, or memorize one equation, $°C = \frac{5}{9} (°F - 32°)$, and learn how to rearrange it so that what was once the independent variable becomes the dependent variable or the unknown quantity. The second alternative is better, for once you have mastered a few algebraic rules (reviewed in the rest of the chapter), you will not have to clutter your memory banks with a lot of duplicated information.

Our calculations for this example would be as follows.

$°C = \frac{5}{9} (°F - 32)$
$°F = \frac{9}{5} °C + 32$
$°F = \frac{9}{5} (0) + 32$
$°F = 32$

Now, consider this example. The mean score on a test equals the sum of the individual scores divided by the number of individuals. Suppose there were three students.

m = mean
x = first student's score
y = second student's score
z = third student's score
thus $m = \dfrac{x + y + z}{3}$
m is the dependent variable
x, y, and z are the independent variables

Now, suppose that the mean was known to be 80 and we knew x and y. We could rearrange the equation and make z the independent variable.

$$3(80) = x + y + z$$
$$240 - x - y = z$$

PROBLEMS. Find the independent and dependent variables for the following equations.

(a) $V = lwh$ _____

(b) $t = \dfrac{2s + 3x + 15}{11}$ _____

(c) $i = prt$ _____

— — — — — — — — — — — — — — — — —

(a) v = dependent variable; l, w, and h = independent variables
(b) t = dependent variable; s and x = independent variables
(c) i = dependent variable; p, r, and t = independent variables

ALGEBRAIC RULES

6 The objective of rearranging or performing operations on equations is to "solve" the equation so that one variable, the unknown, is on one side and all the other terms of the equation are on the other side. To solve equations a set of algebraic rules must be followed. These rules stem from this principle of algebra: *Whatever operation you perform on one side of an equation, you must perform on the other side.*

Rule 1. Whatever is added to one side of an equation must be added to the other.

Examples: Given $X - 2 = Y$, solve for X.
If we add +2 to both sides, we find a solution for X.
$$X - 2 + 2 = Y + 2$$
$$X + 0 = Y + 2$$
$$X = Y + 2$$

Given $X + 2 = Y$, solve for X.
If we add -2 to both sides, we find a solution for X.
$$X + 2 - 2 = Y - 2$$
$$X + 0 = Y - 2$$
$$X = Y - 2$$

PROBLEM. Given $M + B = R$, solve for M.

————————————————

Add $-B$ to both sides.
$M + B - B = R - B$
$M + 0 = R - B$
$M = R - B$

(7) We continue with our rules for solving equations.

Rule 2. Whatever is multiplied through one side of an equation must be multiplied through the other. We do this in order to divide out a factor.

Examples: Given $\dfrac{X}{3} = Y$, solve for X.

Multiply through by 3 and divide the 3's out on the left side.

$\dfrac{X}{3} = Y$

$\dfrac{\cancel{3}X}{\cancel{3}} = 3Y$

$X = 3Y$

Given $D = \dfrac{M}{V}$, solve for M.

Multiply through by V and divide out the V terms on the right hand side.

$DV = \dfrac{M\cancel{V}}{\cancel{V}}$

$DV = M$

Given $D = \dfrac{M}{V}$, solve for V.

Multiply through by V, and divide out the V terms on the right hand side.

$D = \dfrac{M}{V}$

$DV = \dfrac{M\cancel{V}}{\cancel{V}}$

This operation puts the V term in the numerator (that is, gets it out of the fraction). Then multiply by the reciprocal of D and divide out the D terms on the left hand side.

$$\frac{1}{\cancel{D}} \cdot \cancel{D} V = M \times \frac{1}{D}$$

$$V = \frac{M}{D}$$

PROBLEM. Given $3x = 5z$, solve for x.

– – – – – – – – – – – –

Multiply both sides by $\frac{1}{3}$.

$$\frac{1}{\cancel{3}} \cdot \cancel{3}x = \frac{1}{3} \cdot 5z$$

$$x = \frac{5z}{3}$$

⑧   PROBLEM. Given $\frac{ab}{8} = \frac{c}{10}$, solve for b.

– – – – – – – – – – – –

Multiply both sides by $\frac{8}{a}$.

$$\frac{\cancel{8}}{\cancel{a}} \cdot \frac{\cancel{a}b}{\cancel{8}} = \frac{\cancel{8}^{4}}{a} \cdot \frac{c}{\cancel{10}_{5}}$$

$$b = \frac{4c}{5a}$$

⑨   PROBLEM. Given $P = \frac{nRT}{V}$, solve for T.

– – – – – – – – – – – –

Multiply both sides of the equation by V, divide out V terms, and then multiply by $\dfrac{1}{nR}$ and divide out nR terms.

$$P = \frac{nRT}{V}$$

$$VP = \frac{nRT}{V}\,V$$

$$\frac{1}{nR} \cdot VP = nRT\,\frac{1}{nR}$$

$$\frac{VP}{nR} = T$$

SOLVING ALGEBRAIC EQUATIONS

(10)   Some basic tactical approaches can simplify the process of solving equations.

Tactic one. If the variable that we are solving for appears in parentheses, rid the expression of the parentheses.

Example: Given $4(X + 2) = Y$, solve for X.

Either (Method 1) multiply $(X + 2)$ by 4 and then solve in normal fashion or (Method 2) multiply both sides by $\dfrac{1}{4}$ and solve.

Method 1.
$4(X + 2) = Y$
$4X + 8 = Y$
$4X = Y - 8$ (Add -8 to both sides.)
$X = \dfrac{Y - 8}{4}$ or $\dfrac{Y}{4} - 2$ (Multiply both sides by $\dfrac{1}{4}$ ·)

Method 2.
$4(X + 2) = Y$
$X + 2 = \dfrac{Y}{4}$ (Multiply both sides by $\dfrac{1}{4}$.)

$X = \dfrac{Y}{4} - 2$ (Add -2 to both sides.)

Example: Given $2\left(\dfrac{a}{2} + 1\right) = b$, solve for a.

It takes less work to multiply $\dfrac{a}{2} + 1$ by 2 than to multiply the whole left side by $\dfrac{1}{2}$ according to Method 2 above.

$$\frac{\cancel{2}a}{\cancel{2}} + 2 = b \quad \text{(Multiply out, or distribute, on the left side.)}$$

$$a + 2 = b \quad \text{(Divide out 2 in } \frac{2a}{2} \text{.)}$$

$$a = b - 2 \quad \text{(Add } -2 \text{ to both sides.)}$$

Example: Given $M = \dfrac{2}{3}(N - P)$, solve for N.

Multiply by $\dfrac{3}{2}$ then add P.

$$\frac{3}{2}M = \frac{\cancel{3}}{\cancel{2}} \cdot \frac{\cancel{2}}{\cancel{3}}(N - P) \quad \text{(Multiply both sides by } \frac{3}{2} \text{.)}$$

$$\frac{3}{2}M = N - P \quad \text{(Divide out 3 and 2 on the right side.)}$$

$$\frac{3}{2}M + P = N \quad \text{(Add } P \text{ to both sides.)}$$

PROBLEM. Given $°C = \dfrac{5}{9}(°F - 32°)$, solve for $°F$.

— — — — — — — — — — — — —

We considered this equation earlier as we solved for the dependent variable. With your understanding of algebraic tactics, you probably know that we would multiply both sides by $\dfrac{9}{5}$ and then add $32°$ to both sides.

$$°C = \frac{5}{9}(°F - 32°)$$

$$\frac{9}{5}°C = \frac{\cancel{9}}{\cancel{5}} \cdot \frac{\cancel{5}}{\cancel{9}}(°F - 32°)$$

$$\frac{9}{5}°C = °F - 32°$$

$$\frac{9}{5}°C + 32° = °F$$

(11) Tactic two. Where possible, do the addition steps first.

Example: Given $3X + 4 = Y$, solve for X.

Add -4 to each side, then multiply by $\dfrac{1}{3}$.

$$3X + 4 - 4 = Y - 4$$

$$3X = Y - 4$$

$$\frac{3X}{3} = \frac{1}{3}(Y - 4)$$

$$X = \frac{Y - 4}{3} \text{ or } X = \frac{Y}{3} - \frac{4}{3}$$

Example: Given $4 - X = Y$, solve for X.

Add -4 to each side and then multiply by -1. Note that we must solve for X, not $-X$.

$4 - 4 - X = Y - 4$

$-X = Y - 4$

$(-1)(-X) = (-1)(Y - 4)$

$X = -Y + 4$ or $X = 4 - Y$

Example: Given $3(1 - X) = Y$, solve for X.

We cannot add any terms in our attempt to isolate X.

We can: (1) multiply by $\frac{1}{3}$ and then add -1, or (2) carry out the indicated multiplication and solve the equation $3 - 3X = Y$. Method 1 requires fewer steps.

Method 1.

$3(1 - X) = Y$

$\frac{3(1 - X)}{3} = \frac{Y}{3}$

$1 - X = \frac{Y}{3}$

$-X = \frac{Y}{3} - 1$

$X = -\frac{Y}{3} + 1$ or $X = 1 - \frac{Y}{3}$

Method 2.

$3(1 - X) = Y$

$3 - 3X = Y$

$-3X = Y - 3$

$\frac{-3X}{-3} = -\frac{1}{3}(Y - 3)$ (Multiply by $-\frac{1}{3}$.)

$X = -\frac{1}{3}(Y - 3)$

$X = -\frac{Y}{3} + \frac{3}{3}$

$X = -\frac{Y}{3} + 1$ or $X = 1 - \frac{Y}{3}$

PROBLEM. Given $9y - 11 = z$, solve for y.

Add 11 then multiply by $\frac{1}{9}$

$9y - 11 + 11 = z + 11$

$\frac{1}{\cancel{9}} \cdot \cancel{9}y = (z + 11)\frac{1}{9}$

$y = \frac{z + 11}{9}$

(12)   PROBLEM. Given $5(X + 3 - 5) = Y$, solve for X.

— — — — — — — — — — — —

Simplify the expression by adding $+3$ and -5 together, multiply by $\frac{1}{5}$, and add 2.

$5(X - 2) = Y$

$\frac{\cancel{5}(X - 2)}{\cancel{5}} = \frac{Y}{5}$

$X - 2 = \frac{Y}{5}$

$X = \frac{Y}{5} + 2$

(13)   Tactic three. Factoring: Suppose we have the equation $3A + QA = Y + 2$, and we wish to solve for A. On inspection we realize that the 3 and the Q terms cannot be removed easily by multiplication. For example, multiplication by $\frac{1}{3}$ only takes the 3 from the first term and puts it under the second term. In fact, no matter how we play with the terms 3 and Q, there will be no way that we can "free" the A term.

We can, however, multiply through by $\frac{1}{A}$ and solve from there.

$\frac{1}{A}(3A + QA) = \frac{Y + 2}{A}$

$\frac{3\cancel{A}}{\cancel{A}} + \frac{Q\cancel{A}}{\cancel{A}} = \frac{Y + 2}{A}$

$3 + Q = \frac{Y + 2}{A}$

$A(3 + Q) = \frac{\cancel{A}(Y + 2)}{\cancel{A}}$

$A(3 + Q) = (Y + 2)$

$\frac{A\cancel{(3 + Q)}}{\cancel{(3 + Q)}} = \frac{(Y + 2)}{(3 + Q)}$

$$A = \frac{(Y + 2)}{(3 + Q)}$$

That is a lot of work — but there is an easier way. We can *factor out* A. If we notice that $3A + QA$ is the same as $A(3 + Q)$, we have "factored out A." The process of removing a common multiple or factor from a sum of terms is called factoring. We are then in a position to solve the equation quickly.

$A(3 + Q) = Y + 2$ (Factor out A.)

$A\dfrac{\cancel{(3 + Q)}}{\cancel{(3 + Q)}} = \dfrac{(Y + 2)}{(3 + Q)}$ (Multiply both sides by $\dfrac{1}{3 + Q}$ and divide out on the left side.)

$A = \dfrac{Y + 2}{3 + Q}$

Remember that if you're solving for a term that is a multiple of other terms, simplify and factor out the unknown term.

Examples: Given $RX - PX = Y$, solve for X.
Factor out the X, then solve.
$X(R - P) = Y$ (Factor out X.)
$\dfrac{X\cancel{(R - P)}}{\cancel{(R - P)}} = \dfrac{Y}{(R - P)}$ (Multiply both sides by $\dfrac{1}{R - P}$ and divide out on the left side.)
$X = \dfrac{Y}{(R - P)}$

Given $4X + XY + 52 = R$, solve for X.
Factor out X, then solve in the usual manner.
$X(4 + Y) + 52 = R$ (Factor out X.)
$X(4 + Y) = R - 52$ (Add −52 to both sides.)
$\dfrac{X\cancel{(4 + Y)}}{\cancel{(4 + Y)}} = \dfrac{R - 52}{(4 + Y)}$ (Multiply both sides by $\dfrac{1}{4 + Y}$ and divide out on the left side.)
$X = \dfrac{R - 52}{4 + Y}$

Given $P = \dfrac{nRT}{V} - \dfrac{a}{V}$, solve for V.

Factor out $\dfrac{1}{V}$ and then solve as usual.

$P = \dfrac{1}{V}(nRT - a)$ (Factor out $\dfrac{1}{V}$ rather than V because V is in the denominator.)

$VP = \dfrac{\cancel{V}}{\cancel{V}}(nRT - a)$ (Multiply both sides by V.)

$VP = nRT - a$ (Divide out $\dfrac{1}{V}$ on the right side.)

$\dfrac{V\cancel{P}}{\cancel{P}} = \dfrac{1}{P}(nRT - a)$ (Multiply both sides by $\dfrac{1}{P}$ and divide out on the left side.)

$V = \dfrac{nRT - a}{P}$ or $V = \dfrac{nRT}{P} - \dfrac{a}{P}$

PROBLEM. Given $3x + ax = y - 11$, solve for x.

— — — — — — — — — — — — —

Factor out x, then solve.

$x(3 + a) = y - 11$

$\dfrac{x\cancel{(3 + a)}}{\cancel{(3 + a)}} = \dfrac{y - 11}{3 + a}$

$x = \dfrac{y - 11}{3 + a}$

(14)   PROBLEM. Given $H_1M_1(T_f - T_1) = H_2M_2(T_f - T_2)$, solve for T_f. (Hint: Here we have the terms H_1M_1 and H_2M_2 already factored out, but we are solving for T_f term, so we must do the reverse of factoring, that is, multiply out all terms and solve for T_f.)

— — — — — — — — — — — — —

$H_1M_1T_f - H_1M_1T_1 = H_2M_2T_f - H_2M_2T_2$

$H_1M_1T_f = H_2M_2T_f - H_2M_2T_2 + H_1M_1T_1$ (Add $H_1M_1T_1$.)

$H_1M_1T_f - H_2M_2T_f = H_1M_1T_1 - H_2M_2T_2$ (Add $-H_2M_2T_f$.)

Now we have T_f on one side, as a common factor.

$T_f(H_1M_1 - H_2M_2) = H_1M_1T_1 - H_2M_2T_2$

$\dfrac{T_f\cancel{(H_1M_1 - H_2M_2)}}{\cancel{(H_1M_1 - H_2M_2)}} = \dfrac{H_1M_1T_1 - H_2M_2T_2}{(H_1M_1 - H_2M_2)}$ (Multiply by $\dfrac{1}{H_1M_1 - H_2M_2}$.)

$T_f = \dfrac{H_1M_1T_1 - H_2M_2T_2}{H_1M_1 - H_2M_2}$

If you did this correctly, buy yourself a beer!

APPLICATION OF ALGEBRAIC TECHNIQUES
TO PROBLEM SOLVING

(15)   In Chapter Seven we will see that many problems in the sciences can be solved without algebra, but the techniques of algebra are often

useful. If nothing else, learning to think abstractly and relating real quantities to abstract terms will help us understand complicated concepts. With this in mind, let us explore how to solve problems by applying what we have learned of algebra.

Unfortunately, not every problem we encounter will have a "given" equation or one that we have already memorized. When this occurs we must extract enough information from the statement of the problem to construct an algebraic equation relating the variables in such a way that we can solve for the unknown variable. To do this, we should follow the disciplined, step-by-step approach outlined below.

Step 1. Write "Unknown," and note what quantity the problem is asking us to find.

Step 2. Write "Given," and list all the pertinent information furnished in the problem.

Step 3. Assign symbols to the variables.

Step 4. Set up algebraic equations relating the variables.

Step 5. Solve for the unknown variable and then substitute for symbols and calculate.

Example: In most chemistry classes the number of males is three times greater than the number of females. How many females are in a typical class of 400 students?

Step 1. Unknown = number of females.

Step 2. Given = number of males is three times the number of females, and the total number of students = 400.

Step 3. Let X = number of females.
Then number of males = $3X$.
Let S = total number of students.

Step 4. The sum of male and female students will give the total number of students.
$X + 3X = S$
and $4X = S$

Step 5. $4X = S$

$$X = \frac{S}{4}$$

$$X = \frac{400}{4}$$

$X = 100$, the number of females in a class of 400 students.

The following problems may include terms that are unfamiliar to you. If you allow yourself to think abstractly, it should not affect your solutions. If your symbols differ from those given in the solutions, that is perfectly all right. It is the thought and answers that count.

Example: For a series of molecules, it is found that the number of hydrogen atoms exceeds the number of carbon atoms by twice the number of carbon atoms plus two.

How many carbon atoms are in a molecule that has ten hydrogen atoms?

(1) Unknown = number of carbon atoms.

(2) Given = number of hydrogen atoms is equal to twice the number of carbon atoms plus two and is equal to ten.

(3) Let N_C = number of carbon atoms.

Let N_H = number of hydrogen atoms.

(4) $N_H = 2N_C + 2$

(5) $N_H = 2N_C + 2$

$N_H - 2 = 2N_C$

$$\frac{N_H - 2}{2} = N_C$$

$$\frac{10 - 2}{2} = N_C$$

$$\frac{8}{2} = N_C$$

$$4 = N_C$$

PROBLEM. In a particular reaction the amount of oxygen consumed is eight times as great as the amount of hydrogen consumed. How much hydrogen is consumed when 32 grams of oxygen are consumed?

— — — — — — — — — — — — — —

(1) Unknown = number of grams of hydrogen consumed.

(2) Given = eight times as much oxygen as hydrogen; 32 grams of oxygen consumed.

(3) Let O = number of grams of oxygen consumed.

Let H = number of grams of hydrogen consumed.

(4) $8H = O$

$$H = \frac{O}{8}$$

$$H = \frac{32 \text{ grams}}{8}$$

$H = 4$ grams

(16)   PROBLEM. The population of a country is growing at 4% per year. This year's census revealed that the population was 13 million. What was the population last year?

— — — — — — — — — — — — — — —

(1)   Unknown = population last year.
(2)   Given = population grew by 4% and population this year was 13 million.
(3) L = population last year.
        T = population this year.
(4) $T = L + (.04)L$
(5)   Solving for L:
        $13,000,000 = 1.04L$
        $12,500,000 = L$

(17)   PROBLEM. A metal rod expands 0.001 inches for every degree Centigrade increase in temperature. If, on warming, the rod expands from an initial length of 2 inches at 25°C to 2.05 inches, what is the temperature of the rod in degrees Centigrade?

— — — — — — — — — — — — — — —

(1)   Unknown = temperature in °C of warmed rod.

(2)   Given = rod 2 inches long at 25°C and 2.05 inches after warming; increases in length 0.001 inches for every degree Centigrade increase.

(3 & 4) In this problem the steps are mutually dependent, so we combine them. Our thinking should proceed like this: The final length will equal the initial length of the rod plus the amount gained from expansion.

Let L_f = final length in inches.

Let L_i = initial length in inches.

Then, $L_f = L_i$ + amount gained from expansion.

The amount of expansion for a 1°C increase is 0.001 inches. For a 2°C increase in temperature, the increase in length is 0.002 inches. In fact, if we multiply the rate of increase by the temperature change we can calculate how much expansion has occurred.

Let R = rate of expansion in inches per °C (0.001 inches/1°C).

Let T_f = final temperature (unknown).

Let T_i = initial temperature (25°C).

Change in temperature = $T_f - T_i$.

Now we can write: $L_f = L_i + R(T_f - T_i)$

(5) $L_f = L_i + R(T_f - T_i)$

$$L_f - L_i = R(T_f - T_i)$$

$$\frac{L_f - L_i}{R} = T_f - T_i$$

$$\frac{L_f - L_i}{R} + T_i = T_f$$

$$\frac{2.05 \text{ inch} - 2.00 \text{ inch}}{\frac{0.001 \text{ inch}}{1°C}} + 25°C = T_f$$

$$\frac{0.05 \text{ inch}}{\frac{0.001 \text{ inch}}{1°C}} + 25°C = T_f$$

$$\frac{1°C \times 0.05 \text{ inch}}{0.001 \text{ inch}} + 25°C = T_f$$

$$50°C + 25°C = T_f$$

$$75°C = T_f$$

This last problem underlines the importance of concepts we will learn in the chapter dealing with dimensional analysis. A firm understanding of dimensional analysis would have saved us considerable reasoning time in setting up the algebraic equation. However, it would have been very difficult to solve this problem without a knowledge of algebra.

CHAPTER FOUR

Exponents

PRETEST AND OBJECTIVES

This pretest outlines the objectives for Chapter Four. It should help you identify what parts of the chapter, if any, you need to read. Try each of the problems which follow. Then check your answers with the answer key, and follow the directions there.

Objectives and Pretest   **Your Answer**

When you complete this chapter, you will be able to:

1. Express constants and variables in terms of positive exponents, multiplying out constants when indicated.

   Write the following in terms of positive exponents:

   (a) $4 \times 4 \times 4 =$ _____

   (b) $a + 1 =$ _____

   (c) $9 \times 9 \times 9 \times 9 \times 8 \times 8 \times 7 =$ _____

   (d) $abcabcab =$ _____

   (e) $(x + y)(x - 2y)(x - 2y) =$
        (Do not try to remove prentheses.)   _____

   Multiply out the constants and show the variables in terms of positive exponents.

   (f) $2^3 \times 2^2 =$ _____

   (g) $3a \times 3a \times 4a =$ _____

   (h) $(2x)(2x)(3a) =$ _____

   (i) 1^4 _____

   (j) 7^1 _____

2. Express constants and variables in terms of negative exponents, multiplying out constants when indicated.

Write the following in terms of negative exponents.

(a) $\dfrac{1}{3} \times \dfrac{1}{3} =$ _____

(b) $\dfrac{1}{a} =$ _____

(c) $\dfrac{1}{(a+x)} \cdot \dfrac{1}{(a-x)} \cdot \dfrac{2}{(a-x)} =$ _____

(d) $\left(\dfrac{3}{5}\right)^3 =$ _____

(e) $\left(\dfrac{1}{2}\right)^{\frac{1}{3}}$ _____

Multiply out the constants and show the variables in terms of negative exponents.

(f) $\dfrac{1}{4} \times \dfrac{2}{4a} \times \dfrac{8}{a} =$ _____

3. Express roots using fractional exponents or radical signs, as required.

Write the following in terms of fractional exponents, then rewrite each expression using radical signs.

(a) the fifth root of eight _____ _____

(b) the square root of (-4) _____ _____

(c) the sixth root of $2a^2 b^4$ _____ _____

(d) the third root of $a + b$ _____ _____

(e) the nth root of bc^3 _____ _____

(f) the second root of $\dfrac{1}{9}$ _____ _____

(g) the fourth root of x^{-2} _____ _____

(h) the ath root of 3 _____ _____

(i) the ninth root of $(x - y)^7$ _____ _____

4. Multiply and divide like quantities containing exponents; and take roots from or raise to a higher power quantities having positive, negative, or fractional exponents.

   Carry out the indicated operations and keep your answers in terms of exponents.

   (a) $25^3 \times 25^4 =$ _____

   (b) $13^2 \times 13^{-5} =$ _____

   (c) $a^4 (a^{-\frac{1}{2}} + 3a) =$ _____

   (d) $\dfrac{(27)^4 B^7}{B^3} =$ _____

   (e) $\dfrac{B^{10}}{B^{-3}}$ _____

   (f) $\dfrac{(a^4)(-a^6)}{a^{10}} =$ _____

   (g) $(x - y)^7 (x + y)^2 (x + y)^{-9}$ _____

   (h) $9^0 =$ _____

   (i) $(4^2)^{\frac{1}{2}} =$ _____

   (j) $\left(\dfrac{1}{a^2}\right)^{\frac{1}{3}} =$ _____

   (k) $(b^{-\frac{1}{4}})^2 =$ _____

   (l) $(x^{\frac{1}{2}} y^{\frac{1}{3}})^6$ _____

5. Solve equations for a variable given in terms of exponents.

   (a) $K = (Y)(X)^4$, solve for X. _____

   (b) $R = (PQ^{\frac{1}{2}})^3$, solve for Q. _____

   (c) $B^{-3} = M^2$, solve for B. _____

6. Apply the quadratic formula to solve appropriate equations.

   Solve $3x^2 = 2x + 1$ for x (obtain a numerical value for x) _____

Answer Key and Directions

Check your answers with the answer key which follows. If you had difficulty working any of the problems or if you want help with a particular objective, read the indicated portion of Chapter Four.

If you missed any of these problems: **See frames:**

1. (a) 4^3
   (b) $(a + 1)^1$
   (c) $9^4 \times 8^2 \times 7^1$
   (d) $a^3 b^3 c^2$
   (e) $(x + y)^1 (x - 2y)^2$
   (f) 32
   (g) $36a^3$
   (h) $12ax^2$
   (i) 1
   (j) 7 frame 1, page 55

2. (a) 3^{-2}
   (b) a^{-1}
   (c) $2(a + x)^{-1} (a - x)^{-2}$
   (d) $\left(\dfrac{5}{3}\right)^{-3}$
   (e) $(2)^{-\frac{1}{3}}$
   (f) a^{-2} frame 2, page 57

3. (a) $8^{\frac{1}{5}}$; $\sqrt[5]{8}$
   (b) $(-4)^{\frac{1}{2}}$; $\sqrt{-4}$
   (c) $(2a^2 b^4)^{\frac{1}{6}}$; $\sqrt[6]{2a^2 b^4}$
   (d) $(a + b)^{\frac{1}{3}}$; $\sqrt[3]{a + b}$
   (e) $(bc^3)^{\frac{1}{n}}$; $\sqrt[n]{bc^3}$
   (f) $\left(\dfrac{1}{9}\right)^{\frac{1}{2}}$; $\sqrt{\dfrac{1}{9}}$
   (g) $(x^{-2})^{\frac{1}{4}}$; $\sqrt[4]{x^{-2}}$
   (h) $(3)^{\frac{1}{a}}$; $\sqrt[a]{3}$
   (i) $\left((x - y)^7\right)^{\frac{1}{9}}$; $\sqrt[9]{(x - y)^7}$ frame 3, page 58

4. (a) 25^7
   (b) 13^{-3}
   (c) $a^{3\frac{1}{2}} + 3a^5$, or $a^{\frac{7}{2}} + 3a^5$
   (d) $27^4 B^4$, or $(27B)^4$

(e) B^{13}

(f) $-a^0$, or -1

(g) $(x-y)^7 (x+y)^{-7}$

(h) 1

(i) 4

(j) $a^{-\frac{2}{3}}$, or $\dfrac{1}{a^{\frac{2}{3}}}$

(k) $b^{-\frac{1}{2}}$

(l) $x^3 y^2$

frame 4, page 60

5. (a) $X = \left(\dfrac{K}{Y}\right)^{\frac{1}{4}}$

(b) $Q = \left(\dfrac{R}{P^3}\right)^{\frac{2}{3}}$

(c) $B = M^{-\frac{2}{3}}$

frames 5-6, page 63

6. 1 or $-\dfrac{1}{3}$

frames 7-8, page 64

EXPONENTS AS A NOTATION

① An exponent is a means of simplifying an expression that indicates a continued multiplication, such as:

$2 \times 2 \times 2 \times 2.$

This multiplication can be written in what is called *exponential form*, 2^4, where the superscript, 4, tells us how many times the number, 2, appears in the multiplication. The superscript is called the *exponent* or the *power* of 2. For 2^4, we would say "two to the fourth power." The term "power" means exponent while the "fourth" means that the value of the exponent is four.

Since an exponent indicates the number of times a quantity appears in a continuous multiplication, a quantity to the first power means that it appears once. Alternately, $x^1 = x$. We usually do not write the 1 for the first power as it is generally understood. Additionally, we will find in frame 4 that any number to the zero power equals 1.

Examples: $2 \cdot 2 = 2^2$
$(6)(6)(6)(7)(7)(8) = (6^3)(7^2)(8^1)$
$b + 2 = (b + 2)^1$
$xxxxyyzzz = x^4 y^2 z^3$
$(J + 2K)(J + 3K)(J + 2K) = (J + 2K)^2 (J + 3K)^1$
$4 = 4^1$

Sometimes it is more appropriate to express only variables in terms of exponents and to multiply out the constants.

Examples: $3^2 \times 4^1 = (3 \times 3)(4) = 36$
$2b \times b \times 3b = 6b^3$
$(3y)(3ay)(2ab) = 18a^2 b^1 y^2$
$12^1 = 12$

PROBLEMS.

Write the following in terms of positive exponents.

(a) $3 \times 3 \times 3 \times 3 \times 3 =$ _____

(b) $2 \times 2 =$ _____

(c) $10 \times 10 \times 10 \times 3 =$ _____

(d) $a \times a \times a =$ _____

(e) $(y - 2)(y + 2)(y + 2) =$ _____

(f) $b \times a \times b \times a \times b =$ _____

(g) $a + b =$ _____

In the following expressions, write the variables in terms of positive exponents and multiply out the constants.

(h) $2^3 \times 3^2 =$ _____

(i) $(3b \times 3b)(3b \times 3b) =$ _____

(j) $3c \times 3cd \times 2cad =$_____

(k) $1^5 =$ _____

(l) $3^2 \times 10^1 =$ _____

_ _ _ _ _ _ _ _ _ _ _ _ _ _ _

(a) 3^5
(b) 2^2
(c) $3^1 \times 10^3$
(d) a^3
(e) $(y - 2)^1 (y + 2)^2$
(f) $a^2 \times b^3 = a^2 b^3$
(g) $(a + b)^1$
(h) $(2 \times 2 \times 2)(3 \times 3) = 8 \times 9 = 72$
(i) $(3b)^4 = (3 \times 3 \times 3 \times 3)b^4 = 81b^4$
(j) $18a^1 c^3 d^2$
(k) $(1 \times 1 \times 1 \times 1 \times 1) = 1$
(l) $(3 \times 3)(10) = 90$

Remember, the exponent 1 is normally understood rather than written out; we have included it here only for practice.

NEGATIVE EXPONENTS

(2) A negative sign in the exponent indicates the reciprocal of the quantity raised to some power. For example:

$$2^{-3} = \frac{1}{2^3} = \frac{1}{2} \times \frac{1}{2} \times \frac{1}{2}$$

The minus sign means invert the quantity (that is, find the reciprocal) and the superscript 3 indicates a continuous multiplication in which the reciprocal $\frac{1}{2}$ occurs three times.

Examples:

$$\frac{1}{4} \times \frac{1}{4} = \frac{1}{4^2} = 4^{-2}$$

$$\frac{1}{a} \times \frac{1}{a} \times \frac{1}{a} = \frac{1}{a^3} = a^{-3}$$

$$\frac{1}{a-x} = (a-x)^{-1}$$

$$\frac{3}{4} \times \frac{3}{4} = \left(\frac{3}{4}\right)^2 = \frac{1}{\left(\frac{4}{3}\right)^2} = \left(\frac{4}{3}\right)^{-2}$$

$$\left(\text{Note: } \frac{3}{4} = \frac{\cancel{3}}{\frac{4}{\cancel{3}}} = \frac{1}{\frac{4}{3}}.\right)$$

$$\frac{1}{5^{\frac{1}{2}}} = 5^{-\frac{1}{2}}$$

PROBLEMS. Write the following in terms of negative exponents.

(a) $\dfrac{1}{3} \times \dfrac{1}{3} =$ _____

(b) $\dfrac{1}{10} \times \dfrac{1}{10} \times \dfrac{1}{10} \times \dfrac{1}{2} \times \dfrac{1}{2} =$ _____

(c) $\dfrac{1}{(a+x)} \cdot \dfrac{1}{(a+x)} \cdot \dfrac{2}{(a+x)} \cdot \dfrac{1}{(a-x)} =$ _____

(d) $\dfrac{1}{b+c} =$ _____

(e) $\dfrac{1}{3b} \times \dfrac{2}{4b} =$ _____

(f) $\dfrac{1}{ab} \times \dfrac{1}{bc} \times \dfrac{1}{a} \times 2 =$ _____

(g) $\left(\dfrac{8}{9}\right)^{6} =$ _____

(h) $\left(\dfrac{1}{4}\right)^{\frac{1}{5}} =$ _____

- - - - - - - - - - - - -

(a) 3^{-2}

(b) $2^{-2} \times 10^{-3}$

(c) $\dfrac{2}{(a + x)^{3}(a - x)} = 2(a - x)^{-1}(a + x)^{-3}$

(d) $(b + c)^{-1}$

(e) $\dfrac{2}{12b^{2}} = \dfrac{1}{6b^{2}} = \dfrac{1}{6}b^{-2.} = 6^{-1}b^{-2}$

(f) $\dfrac{2}{a^{2}b^{2}c} = 2a^{-2}b^{-2}c^{-1}$

(g) $\left(\dfrac{9}{8}\right)^{-6}$

(h) $(4)^{-\frac{1}{5}}$

ROOTS AND FRACTIONAL EXPONENTS

③   The reverse of raising a quantity to a power is called finding a *root* of that quantity. It means, for example, that if we are taking the third root of a number like 64, we are looking for the number that appears in a continuous multiplication three times to give a value of 64.

third root of 64 = 4; $4^{3} = 64$

To indicate the root of a number we often use the radical sign $\sqrt{}$. ("Radical" means root.) The notation $\sqrt[n]{A}$ means that we are seeking the nth root of A. In the example below we seek some quantity, Y, that appears in a continuous multiplication n times to give A.

$\sqrt[n]{A} = Y;\quad Y^{n} = A$

$\sqrt[3]{8} = 2;\quad 2^{3} = 8$

Generally, if $n = 2$, the 2 is not written with the radical sign, since it is understood that the radical signifies the second root, or square root, unless other roots are indicated.

Examples: the second root of $4 = \sqrt{4}$

the third root of $27 = \sqrt[3]{27}$

the fourth root of $10,000 = \sqrt[4]{10,000}$

the fourth root of $1 = \sqrt[4]{1}$

the second root of $a \cdot a = \sqrt{a \cdot a}$ or $\sqrt{a^2}$

the third root of $(a + b)^3 = \sqrt[3]{(a + b)^3}$

the second root of $\frac{1}{4} = \sqrt{\frac{1}{4}}$

the nth root of $xyz = \sqrt[n]{xyz}$

the fifth root of $30^{-3} = \sqrt[5]{30^{-3}}$ or $\sqrt[5]{1/30^3}$

the xth root of $2 = \sqrt[x]{2}$

(Note: x must be a natural number, e.g., 1, 2, 3 ....)

Often it is more convenient to express roots as a fractional exponent with the reciprocal of the root (that is, 1 over the root) to be taken as the exponent.

Examples: $\sqrt{4} = 4^{\frac{1}{2}}$

$\sqrt[3]{27} = 27^{\frac{1}{3}}$

$\sqrt[4]{10,000} = (10,000)^{\frac{1}{4}}$

$\sqrt[4]{1} = 1^{\frac{1}{4}} = 1$

$\sqrt[2]{a \cdot a} = (a \cdot a)^{\frac{1}{2}} = (a^2)^{\frac{1}{2}}$

$\sqrt[3]{(a + b)^3} = ((a + b)^3)^{\frac{1}{3}}$

$\sqrt{\frac{1}{4}} = (\frac{1}{4})^{\frac{1}{2}} = (4^{-1})^{\frac{1}{2}}$

(We'll see in the next frame how to further simplify these exponents.)

$\sqrt[n]{xyz} = (xyz)^{\frac{1}{n}}$

$\sqrt[5]{1/30^3} = (1/30^3)^{\frac{1}{5}}$

$\sqrt[x]{2} = 2^{\frac{1}{x}}$

PROBLEMS. Express the following as fractional exponents, then rewrite each expression using a radical sign.

(a) the fourth root of $16 =$ _____ = _____

(b) the third root of $8 =$ _____ = _____

(c) the ath root of $100 =$ _____ = _____

(d) the fifth root of $3^{-4} =$ _____ = _____

(e) the third root of $9y =$ _____ = _____

(f) the fifth root of $\frac{1}{32} =$ _____ = _____

(g)   the nth root of $24x^2$ _____ = _____

(h)   the sixth root of $(a + b)^7$ _____ = _____

- - - - - - - - - - - - - - -

(a) $16^{\frac{1}{4}}$; $\sqrt[4]{16}$

(b) $8^{\frac{1}{3}}$; $\sqrt[3]{8}$

(c) $100^{\frac{1}{a}}$; $\sqrt[a]{100}$

(d) $(3^{-4})^{\frac{1}{5}}$; $\sqrt[5]{3^{-4}}$

(e) $(9y)^{\frac{1}{3}}$; $\sqrt[3]{9y}$

(f) $\left(\dfrac{1}{32}\right)^{\frac{1}{5}} = (32^{-1})^{\frac{1}{5}}$; $\sqrt[5]{\dfrac{1}{32}}$

(g) $(24x^2)^{\frac{1}{n}}$; $\sqrt[n]{24x^2}$

(h) $\left((a + b)^7\right)^{\frac{1}{6}}$; $\sqrt[6]{(a + b)^7}$

MATH OPERATIONS ON EXPONENTS

4   Like numbers, exponents can be multiplied, subtracted, and added. When they are, they obey the same laws for these operations as discussed in previous chapters. However, since they are exponents and not numbers per se, we follow special rules for operating on them.

Rule 1. When multiplying *like* quantities we add exponents.

Examples: $2 \times 2 = 2^{1 + 1} = 2^2$

$4^2 \times 4^{-1} = 4^{2 - 1} = 4^1$

$(a + b)^3 (a + b)^{-4} = (a + b)^{3 - 4} = (a + b)^{-1}$

(Note: $(a + b)^{-1} \neq a^{-1} + b^{-1}$. One cannot distribute an exponent across addition or subtraction.)

$\sqrt{9 + 16} = \sqrt{25} = 5$, but $\sqrt{9 + 16} = (9 + 16)^{\frac{1}{2}}$ $\neq 9^{\frac{1}{2}} + 16^{\frac{1}{2}}$ because $\sqrt{9} + \sqrt{16} = 3 + 4 = 7$.)

$2^3 \times 3^4 = 2^3 \times 3^4$ (Must be like quantities to add exponents.)

$10^4 \times 10^{\frac{1}{2}} = 10^{4 + \frac{1}{2}} = 10^{4\frac{1}{2}} = 10^{\frac{9}{2}}$

$(a^2 b^3)(a^4 b^4) = a^{2 + 4} b^{3 + 4} = a^6 b^7$

(Note: $a^4 b^4 = (ab)^4$. One can distribute an exponent across multiplication or division since the exponent involves repeated multiplication anyway.)

$(x^5)(-x)^2 = x^5 (-1)^2 x^2 = x^7$, since $(-1)^2 = (-1)(-1) = 1$

(Note: $(-x)^2 \neq (-x^2)$, since $(-x^2)$ does not involve squaring -1.)

$a^2 (a + a^5) = a^2 (a) + a^2 (a^5) = a^3 + a^7$

Rule 2. When dividing like quantities we subtract the exponent of the quantity in the denominator from the exponent of the *same* quantity in the numerator (or vice versa if we want our answer in the denominator instead)

Examples: $\dfrac{3^2}{3^5} = 3^{2-5} = 3^{-3}$

(Note: Same as $3^2 \times 3^{-5} = 3^{-3}$.)

$\dfrac{a^2 b^6}{a^1 b^2 c} = a^{2-1} b^{6-2} c^{-1} = ab^4 c^{-1}$

$\dfrac{4^3}{4^{½}} = 4^{3-\frac{1}{2}} = 4^{2\frac{1}{2}} = 4^{\frac{5}{2}}$

$\dfrac{10^6}{3 \times 10^4} = \dfrac{1}{3} \times 10^{6-4} = \dfrac{1}{3} \times 10^2$

$\dfrac{10^{-5}}{10^3} = 10^{-5-3} = 10^{-8}$ or $\dfrac{1}{10^{3-(-5)}} = \dfrac{1}{10^8}$

$\dfrac{a^2}{a^{-b}} = a^{2-(-b)} = a^{2+b}$

Rule 3. Any quantity ($\neq 0$) raised to the zero power is equal to 1.

$\dfrac{10^2}{10^2} = 1 = \dfrac{10^2}{10^2} = 10^{2-2} = 10^0$

Examples: $a^0 = 1$
$356^0 = 1$
$\dfrac{(3a^2)^2}{(3a)^2} = \dfrac{3^2 a^4}{3^2 a^2} = 3^{2-2} a^{4-2} = a^2$
$(3x^2 + 7x)^0 = 1$

Rule 4. To take the root of a quantity that is raised to a power, we multiply the exponent of that quantity by the root taken. To raise a quantity with a fractional exponent to a higher power we multiply the power and fractional exponent.

Examples: $\sqrt{2^2} = (2^2)^{\frac{1}{2}} = 2^{2 \times \frac{1}{2}} = 2^{\frac{2}{2}} = 2^1$
$(a^3)^{\frac{1}{3}} = a^{3 \times \frac{1}{3}} = a^{\frac{3}{3}} = a$
$(M^5)^{\frac{1}{3}} = M^{\frac{1}{3} \times 5} = M^{\frac{5}{3}}$
$(\sqrt{2})^2 = (2^{\frac{1}{2}})^2 = 2^{2 \times \frac{1}{2}} = 2^{\frac{2}{2}} = 2^1$
$(a^{\frac{1}{3}})^3 = a^{3 \times \frac{1}{3}} = a^{\frac{3}{3}} = a$
$(M^{\frac{1}{3}})^5 = M^{\frac{1}{3} \times 5} = M^{\frac{5}{3}}$

PROBLEMS. Carry out the indicated operations and keep answers in exponential notation where possible.

(a) $\dfrac{10^6 \times 10^5}{10^9} =$ _____

(b) $\dfrac{2^2 \times 3^3}{2 \times 3} =$ _____

(c) $\dfrac{a^b}{a^c M} =$ _____

(d) $\dfrac{a(a^2 + b)}{10^2} =$ _____

(e) $\dfrac{M^2}{M^{\frac{1}{3}}} =$ _____

(f) $(M^{\frac{1}{3}})(M^{\frac{1}{4}}) =$ _____

(g) $\dfrac{Q^5}{Q^3 \cdot Q^2} =$ _____

(h) $\dfrac{1 \times 10^4}{1 \times 10^{-6}} =$ _____

(i) $\dfrac{x^5}{x^{\frac{1}{2}}} =$ _____

(j) $x^6(-xy^6) =$ _____

(k) $\left((a + b)^2\right)^{\frac{1}{4}} =$ _____

(l) $\dfrac{(x + y)^7}{(x - y)^2(x + y)^2} =$ _____

(m) $(a^{\frac{1}{2}} b^{\frac{1}{4}})^8 =$ _____

(n) $\left((7x^2)^{\frac{1}{2}}\right)^0 =$ _____

– – – – – – – – – – – – – – –

(a) $10^{6+5-9} = 10^{11-9} = 10^2$

(b) $2^{2-1} \times 3^{3-1} = 2 \times 3^2$

(c) $a^{b-c} \cdot M^{-1}$

(d) $\dfrac{a^{2+1} + ab}{10^2} = \dfrac{a^3 + ab}{10^2} = (a^3 + ab)10^{-2}$

(e) $M^{2-(\frac{1}{3})} = M^{2+\frac{1}{3}} = M^{2\frac{1}{3}}$

(f) $M^{\frac{1}{3}+\frac{1}{4}} = M^{\frac{7}{12}}$

(g) $Q^{5-(3+2)} = Q^{5-5} = Q^0 = 1$

(h) $1 \times 10^{4-(-6)} = 1 \times 10^{10}$

(i) $x^{5-\frac{1}{2}} = x^{4\frac{1}{2}} = x^{\frac{9}{2}}$

(j) $x^{6+1}y^6(-1) = -x^7y^6$

(k) $(a + b)^{2 \cdot \frac{1}{4}} = (a + b)^{\frac{1}{2}}$

(l) $\dfrac{(x + y)^{7-2}}{(x - y)^2} = \dfrac{(x + y)^5}{(x - y)^2}$

(m) $a^{\frac{1}{2} \cdot 8} b^{\frac{1}{4} \cdot 8} = a^4 b^2$

(n) 1

APPLICATION OF EXPONENTS TO PROBLEMS

⑤   Sometimes we find ourselves faced with an equation of the form $x^3 = 8$, and we wish to solve for x. To do so we have to take the third or cube root of each side of the equation.

$$x^3 = 8$$
$$(x^3)^{\frac{1}{3}} = 8^{\frac{1}{3}}$$
$$x^{3 \times \frac{1}{3}} = 8^{\frac{1}{3}}$$
$$x^{\frac{3}{3}} = 8^{\frac{1}{3}}$$
$$x^1 = 8^{\frac{1}{3}}$$

At this point we can find a value for x by consulting the appropriate tables or by logarithms as outlined in Chapter Six.

The point to realize is that $x^3 = 8$ does not constitute a solution of x but for x^3.

Example: Given $x^2 + 5 = y$, solve for x.

$x^2 = y - 5$ (Add -5 to both sides.)

$(x^2)^{\frac{1}{2}} = (y - 5)^{\frac{1}{2}}$ (Take square root of both sides.)

$x^{2 \times \frac{1}{2}} = (y - 5)^{\frac{1}{2}}$

$x = (y - 5)^{\frac{1}{2}}$

Example: Given $K_{SP} = (S)(3S)^3$, solve for S.

$K_{SP} = (S)^1(3)^3(S)^3$

(Note: Because the term is $3 \cdot S$, we can cube both 3 and S. But if the term were $(3 + S)^3$, we could not, because that equals $(3 + S)(3 + S)(3 + S)$.)

$K_{SP} = 27S^4$ (Add exponents of $S - 3 + 1 = 4$, and $3^3 = 27$.)

$\dfrac{K_{SP}}{27} = S^4$ (Multiply both sides by $\dfrac{1}{27}$.)

$\left(\dfrac{K_{SP}}{27}\right)^{\frac{1}{4}} = (S^4)^{\frac{1}{4}}$ (Take 4th root of both sides.)

$\left(\dfrac{K_{SP}}{27}\right)^{\frac{1}{4}} = S$

PROBLEM. Given $y - 7 = x^3 + 10$, solve for x.

—————————————————

$y - 7 - 10 = x^3 + 10 - 10$
$(y - 17)^{\frac{1}{3}} = (x^3)^{\frac{1}{3}}$
$(y - 17)^{\frac{1}{3}} = x$

⑥   PROBLEM. The volume of a cube is defined as the third power of the length of one side of the cube. What is the length in feet of the side of a cube that occupies 125 cubic feet? (Hint: Use the problem-solving steps from Chapter Three, frame 15.)

—————————————————

(1)   Unknown: length of a side of a cube in feet.
(2)   Given: cube = 125 cubic feet; volume = length of side of cube raised to the third power.
(3)   Let V = volume.
      Let L = length of side of cube.
(4) $V = L^3$
(5) $V^{\frac{1}{3}} = (L^3)^{\frac{1}{3}}$
      $V^{\frac{1}{3}} = L$
      $(125 \text{ cubic feet})^{\frac{1}{3}} = L$
      5 feet = L

THE QUADRATIC FORMULA

⑦   At some time during your first year of college science, you may come in contact with the following type of equation, $ax^2 + bx + c = 0$.

In this equation a, b, and c are constants; x^2 and x are variables. Be prepared to see this equation in several forms and to rearrange it into the standard form as shown above.

Examples: $x^2 = 2x + 1$ (Rearrange by adding $-2x$ and -1 to both sides: $x^2 - 2x - 1 = 0$.)

$x(x - 1) = 3$ (Rearrange by multiplying out $x(x - 1)$ and adding -3 to both sides: $x^2 - x - 3 = 0$.)

The key point to remember is that the variable is raised to the second power (multiplied by itself) in a quadratic equation.

One way to solve a quadratic equation is to use the quadratic formula:

$$x = \frac{-b \pm \sqrt{b^2 - 4ac}}{2a}$$

You simply plug in the appropriate constants to find x, arriving at two solutions:

$$\frac{-b + \sqrt{b^2 - 4ac}}{2a}$$

$$\frac{-b - \sqrt{b^2 - 4ac}}{2a}$$

Example: Given $2x^2 = 3x + 1$, solve for x.

Since the quadratic formula is used for the quadratic equation in the form $ax^2 + bx + c = 0$, it is important to rearrange the equation so that all terms are on one side. We add $-3x$ and -1 to both sides and obtain:

$2x^2 - 3x - 1 = 3x + 1 - 3x - 1 = 0$

Now we can apply the quadratic formula, substituting into it the values, $a = 2$, $b = -3$, $c = -1$.

$$x = \frac{-(-3) \pm \sqrt{(-3)^2 - 4(2)(-1)}}{2(2)}$$

$$x = \frac{3 \pm \sqrt{9 + 8}}{4}$$

$$x = \frac{3 + \sqrt{17}}{4} \text{ and } x = \frac{3 - \sqrt{17}}{4}$$

If you wish, you can consult a chart to find the $\sqrt{17}$. However, the above answer is mathematically correct.

Example: Given $K = \dfrac{x^2}{M - x}$, solve for x.

This is the ionization equation in chemistry.
K = a specific constant for each kind of substance that can be dissolved and ionized.

$M =$ a constant which is the amount of substance that has been dissolved.

$x =$ the unknown which is the amount of dissolved substance that has been ionized.

To solve for x, we first arrange the equation in the quadratic equation's standard form. We multiply both sides by $(M - x)$ and obtain:

$$K(M - x) = \frac{x^2}{\cancel{(M - x)}} \;\; \cancel{(M - x)}$$

We then add $-K(M - x)$ to both sides and obtain:

$$K(M - x) - K(M - x) = x - K(M - x)$$
$$0 = x - KM + Kx$$
$$0 = x + Kx - KM$$

Now, apply the quadratic formula, using the values $a = 1$, $b = K$, $c = -KM$:

$$x = \frac{-K \pm \sqrt{K^2 - 4(1)(-KM)}}{2(1)}$$

$$x = \frac{-K \pm \sqrt{K^2 + 4KM}}{2}$$

(Note: $x = \dfrac{-K - \sqrt{K^2 + 4KM}}{2}$ would not be a practical solution because it would indicate a negative amount of substance ionized. Therefore, $x = \dfrac{-K + \sqrt{K^2 + 4KM}}{2}$ would be the solution.)

PROBLEM. Given $x^2 = 4x + 16$, solve for x.

Rewrite the equation to fit the quadratic formula:

$x^2 = 4x + 16$

$x^2 - 4x - 16 = 0$

Apply the quadratic formula: $x = \dfrac{-b \pm \sqrt{b^2 - 4ac}}{2a}$

   Note: $a = 1$, $b = -4$, and $c = -16$

$x = \dfrac{-(-4) \pm \sqrt{(-4)^2 - 4 \times 1 \times (-16)}}{2 \times 1}$

$x = \dfrac{4 \pm \sqrt{16 + 64}}{2}$

$x = \dfrac{4 \pm \sqrt{80}}{2}$

⑧   PROBLEM. Given $2 - 2x^2 = 8x$, solve for x.

– – – – – – – – – – – – – –

Multiply by $\dfrac{1}{2}$ on both sides to make computations easier.

$\dfrac{2 - 2x^2}{2} = \dfrac{8x}{2}$

$1 - x^2 = 4x$

Rewrite the equation to fit the quadratic formula $-x^2 - 4x + 1 = 0$ (add $-4x$ to both sides).

Substituting $a = -1$, $b = -4$, $c = 1$:

$x = \dfrac{-(-4) \pm \sqrt{(-4)^2 - 4(-1)(1)}}{2(-1)}$

$x = \dfrac{4 \pm \sqrt{16 + 4}}{-2}$

$x = \dfrac{4 \pm \sqrt{20}}{-2}$

CHAPTER FIVE
Scientific Notation

PRETEST AND OBJECTIVES

This pretest outlines the objectives for Chapter Five. It should help you identify what parts of the chapter, if any, you need to read. Try each of the problems which follow. Then check your answers with the answer key, and follow the directions given there.

Objectives and Pretest   **Your Answer**

When you complete this chapter, you will be able to:

1. Express given numbers in scientific notation.

    (a)   2,500   _____

    (b)   30   _____

    (c) $\dfrac{1}{100}$ _____

    (d)   0.00006   _____

2. Given numbers in scientific notation, express them in nonexponential form.

    (a) 4.9×10^4 _____

    (b) 5×10^1 _____

    (c) 1×10^{-1} _____

    (d) 6.8×10^{-6} _____

3. Carry out the indicated operations, keeping your answers in scientific notation.

    (a) $5 \times 10^2 + 2 \times 10^4 =$ _____

(b) $3 \times 10^{-1} + 4 \times 10^{1} =$ _____

(c) $4.1 \times 10^{-4} - 2 \times 10^{-5} =$ _____

(d) $6.8 \times 10^{40} - 6 \times 10^{41} =$ _____

4. Carry out the indicated multiplications, keeping your answers in scientific notation.

(a) $(5 \times 10^{3})(6 \times 10^{5}) =$ _____

(b) $(3 \times 10^{-4})(2 \times 10^{8}) =$ _____

(c) $(-4 \times 10^{-8})(-2 \times 10^{-13}) =$ _____

(d) $(1.2 \times 10^{-10})(3 \times 10^{10}) =$ _____

5. Carry out the indicated divisions, keeping your answers in scientific notation.

(a) $\dfrac{4 \times 10^{4}}{2 \times 10^{2}} =$

_____

(b) $\dfrac{3.2 \times 10^{5}}{8 \times 10^{-3}} =$

_____

(c) $\dfrac{8.1 \times 10^{-4}}{9.0 \times 10^{2}} =$

_____

(d) $\dfrac{5.4 \times 10^{-10}}{9 \times 10^{-5}} =$

_____

6. Find the roots of the following, keeping your answers in scientific notation.

(a) $(4 \times 10^{2})^{\frac{1}{2}} =$ _____

(b) $(16 \times 10^{8})^{\frac{1}{4}} =$ _____

(c) $(1.6 \times 10^{-13})^{\frac{1}{2}} =$ _____

7. Raise the following to the indicated powers, keeping your answers in scientific notation.

(a) $(1.1 \times 10^{-3})^{2} =$ _____

(b) $(6 \times 10^{3})^{2} =$ _____

Answer Key and Directions

Check your answers with the answer key which follows. If you had difficulty working any of the problems or if you want help with a particular objective, read the indicated portion of Chapter Five.

If you missed any of these problems: **See frames:**

1. (a) 2.5×10^3
   (b) 3×10^1
   (c) 1×10^{-2}
   (d) 6×10^{-5} frame 1, page 70

2. (a)   49,000
   (b)   50
   (c)   0.1
   (d)   .0000068   frame 2, page 71

3. (a) 2.05×10^4
   (b) 4.03×10^1
   (c) 3.9×10^{-4}
   (d) -5.32×10^{41} frame 3, page 72

4. (a) 3×10^9
   (b) 6×10^4
   (c) 8×10^{-21}
   (d) $3.6 \times 10^0 = 3.6$ frame 4, page 73

5. (a) 2×10^2
   (b) 4×10^7
   (c) 9×10^{-7}
   (d) 6×10^{-6} frame 5, page 74

6. (a) 2×10^1
   (b) 2×10^2
   (c) 4×10^{-7} frame 6, page 75

7. (a) 1.21×10^{-6}
   (b) 3.6×10^7 frame 7, page 75

EXPRESSING NUMBERS

1   In science courses we often encounter numbers that are very large or very small. To avoid the cumbersome task of writing numbers like 6,000,000, and small numbers like 0.000005, we use a systematic presentation called *scientific notation*. Scientific notation is also of great use in our understanding of logarithms.

To review, remember that any number (other than zero) can be expressed as the product of that number and 10^0 (because $10^0 = 1$).

Examples: $3,500 = 3,500 \times 10^0$

$0.00005 = 0.00005 \times 10^0$

To express a number in scientific notation, we rewrite it as the product of a number (from 1 to 9,99...) and a power of 10. The power of 10 tells us how we shifted the decimal point in the first number to get the number we want.

Rule 1. For every time we shift the decimal point to the left we increase the value of the exponent of 10 by one.

Examples: $3,500. = 3.5 \times 10^3$ (The decimal was moved to the left three times.)

$2,480,000,000. = 2.48 \times 10^9$

$70. = 7. \times 10^1$

$435.6 = 4.356 \times 10^2$

$5.47 = 5.47 \times 10^0$
(Note: We did not need to move the decimal so our power of 10 is 10^0, which equals 1.)

Rule 2. For every time we shift the decimal point to the right the exponent of 10 is reduced by one.

Examples: $0.00005 = 5 \times 10^{-5}$ (The decimal was moved to the right five times.)

$0.00043 = 4.3 \times 10^{-4}$ (Notice that we stop shifting when there is one number to the left of the decimal —in this case, the number 4.)

$0.1 = 1 \times 10^{-1}$

PROBLEMS. Express the following in scientific notation.

(a) $3,400 =$ _____   (d) $1/200 =$ _____

(b) $20 =$ _____   (e) $0.00000489 =$ _____

(c) $6,000,000 =$ _____   (f) $2.1 =$ _____

_____

(a) 3.4×10^3; (b) 2×10^1; (c) 6×10^6; (d) 5×10^{-3}; (e) 4.89×10^{-6}; (f) 2.1×10^0

CONVERTING TO NUMBERS

2   To express a number that is written in scientific notation as a non-exponential quantity, we apply the following rules.

Rule 1. If the exponent of 10 is positive, move the decimal to the right the same number of spaces as the value of the exponent, and fill in with zeroes.

Examples: $3.8 \times 10^6 = 3,800,000$

$3 \times 10^2 = 300$

Rule 2. If the exponent of 10 is negative, move the decimal to the left the same number of spaces as the value of the exponent, and fill in with zeroes.

Examples: $4 \times 10^{-5} = 0.00004$

$2 \times 10^{-1} = 0.2$

PROBLEMS. Convert the following into nonexponential numbers.

(a) $6.2 \times 10^5 =$ _____

(c) $4.8 \times 10^{-3} =$ _____

(b) $3.0 \times 10^1 =$ _____

(d) $1 \times 10^{-6} =$ _____

– – – – – – – – – – – – – – –

(a) 620,000;   (b) 30;   (c) 0.0048;   (d) 0.000001

ADDITION AND SUBTRACTION

③   Occasionally, we wish to find the sum of numbers expressed in scientific notation that have different values for the exponent of 10. Because we can only add or subtract *like* quantities, we must first rewrite the numbers so that they have the same exponent of 10. (When rewriting quantities in scientific notation, always remember that if we make the exponent larger we must make the number part smaller, and if we make the exponent smaller we have to make the number part larger.) Then add or subtract the number parts only, carrying the power of 10 unchanged.

Examples: $6 \times 10^2 + 5 \times 10^3 = 0.6 \times 10^3 + 5 \times 10^3 = 5.6 \times 10^3$

or $6 \times 10^2 + 5 \times 10^3 = 6 \times 10^2 + 50 \times 10^2 = 56 \times 10^2$
$= 5.6 \times 10^3$
(Note: 56×10^2 is not standard form, so we must convert the final answer to 5.6×10^3. It is usually more convenient to convert to the largest exponent, because the answer comes out in proper scientific notation without further conversion.)

$4 \times 10^{-2} + 6 \times 10^{-4} = 4 \times 10^{-2} + 0.06 \times 10^{-2} = 4.06 \times 10^{-2}$
(Note: -2 is a larger exponent than -4 so we convert $6. \times 10^{-4}$ to $.06 \times 10^{-2}$. We increased the power of 10 by 2 so we decrease 6 by moving the decimal back (to the left) two places.)

$5.0 \times 10^6 - 6 \times 10^5 = 5.0 \times 10^6 - 0.6 \times 10^6 = 4.4 \times 10^6$
(Note: We increased the power of 10 in $6. \times 10^5$ by 1 so we moved the decimal back one place to $.6 \times 10^6$.)

$4 \times 10^3 - 8 \times 10^4 = 0.4 \times 10^4 - 8 \times 10^4 = -7.6 \times 10^4$
(Note: Once both numbers are at 10^4, we just subtract $.4 - 8 = -7.6$, carrying 10^4 unchanged.)

$3.0 \times 10^{-17} - 9.0 \times 10^{-16} = 0.3 \times 10^{-16} - 9.0 \times 10^{-16}$
$= -8.7 \times 10^{-16}$

PROBLEMS. Carry out the indicated operations.

(a) $2 \times 10^2 + 3 \times 10^4 = $ _____

(b) $1 \times 10^{-1} + 1 \times 10^1 = $ _____

(c) $6 \times 10^{-4} - 3 \times 10^{-5} = $ _____

(d) $2.5 \times 10^{20} - 6 \times 10^{21} = $ _____

– – – – – – – – – – – – – – –

(a) $0.02 \times 10^4 + 3 \times 10^4 = 3.02 \times 10^4$
(b) $0.01 \times 10^1 + 1 \times 10^1 = 1.01 \times 10^1$
(c) $6 \times 10^{-4} - 0.3 \times 10^{-4} = 5.7 \times 10^{-4}$
(d) $0.25 \times 10^{21} - 6 \times 10^{21} = -5.75 \times 10^{21}$

MULTIPLICATION

④   To multiply two quantities expressed in scientific notation, multiply the two values in the number part and add the exponents of 10. If necessary, convert the final answer to the standard form.

Examples: $(6 \times 10^5)(4 \times 10^3) = 6 \times 4 \times 10^{5+3} = 24 \times 10^8$
$= 2.4 \times 10^9$

$(2 \times 10^6)(3 \times 10^{-5}) = 2 \times 3 \times 10^{6+(-5)} = 6 \times 10^1$

$(8.1 \times 10^{-5})(2 \times 10^{-5}) = 16.2 \times 10^{(-5)+(-5)} = 16.2 \times 10^{-10}$
$= 1.62 \times 10^{-9}$

$(4 \times 10^{-2})(2 \times 10^2) = 8 \times 10^{2+(-2)} = 8 \times 10^0 = 8$

PROBLEMS. Carry out the indicated operations.

(a) $(2 \times 10^2)(4 \times 10^{20}) = $ _____

(b) $(1 \times 10^{-5})(-6 \times 10^7) = $ _____

(c) $(4 \times 10^{-8})(4 \times 10^{-20}) = $ _____

(d) $(1.6 \times 10^{-4})(4 \times 10^4) =$ _____

— — — — — — — — — — — — — — —

(a) 8×10^{22}; (b) -6×10^2; (c) 1.6×10^{-27}; (d) $6.4 \times 10^0 = 6.4$

DIVISION

5 To divide in scientific notation, divide the nonexponential numbers in the normal fashion. Then change the sign of the exponent of the 10 in the denominator (divisor) and add it to the exponent of 10 in the numerator (dividend). Convert the final answer to standard notation, if necessary.

Note that when you change the sign in the denominator and add, you are following rule 2 (frame 4, Chapter Four) which says to subtract exponents when dividing. Subtraction means changing the sign and adding (rule 3 of frame 1, Chapter One).

Examples: $\dfrac{3 \times 10^5}{2 \times 10^3} = \dfrac{3}{2} \times 10^{5-3} = 1.5 \times 10^2$

$\dfrac{1.6 \times 10^4}{8 \times 10^{-2}} = \dfrac{1.6}{8} \times 10^{4+2} = 0.2 \times 10^6 = 2 \times 10^5$

$\dfrac{2.5 \times 10^{-8}}{5 \times 10^6} = \dfrac{2.5}{5} \times 10^{-8-6} = 0.5 \times 10^{-14} = 5 \times 10^{-15}$

$\dfrac{6.6 \times 10^{-52}}{3 \times 10^{-6}} = \dfrac{6.6}{3} \times 10^{-52+6} = 2.2 \times 10^{-46}$

PROBLEMS. Carry out the indicated operations.

(a) $\dfrac{4 \times 10^{18}}{2 \times 10^2} =$ _____

(b) $\dfrac{3.2 \times 10^6}{4 \times 10^{-5}} =$ _____

(c) $\dfrac{9 \times 10^{-3}}{3 \times 10^5} =$ _____

(d) $\dfrac{8.1 \times 10^{-10}}{9 \times 10^{-10}} =$ _____

— — — — — — — — — — — — — — —

(a) 2×10^{16}; (b) 8×10^{10}; (c) 3×10^{-8}; (d) 9×10^{-1}

TAKING ROOTS

(6) To find the root of a quantity expressed in scientific notation, take the roots of the nonexponential part and the exponential part separately and multiply the resultant quantities. If necessary, convert the final answer to standard notation form.

Examples:

$$(4 \times 10^{10})^{\frac{1}{2}} = 4^{\frac{1}{2}} \times 10^{\frac{10}{2}} = 2 \times 10^5$$

$$(9 \times 10^{-6})^{\frac{1}{2}} = 9^{\frac{1}{2}} \times 10^{-\frac{6}{2}} = 3 \times 10^{-3}$$

If the fractional exponent and the exponent of 10 does not translate into a whole number, it is generally convenient to rewrite the expression so that the exponent of 10 can be divided evenly by the value of the root, giving a whole number.

Examples:

$$(4 \times 10^{11})^{\frac{1}{2}} = 4^{\frac{1}{2}} \times 10^{\frac{11}{2}} = 2 \times 10^{5.5}$$

Because this is awkward we change it as follows:
$$(4 \times 10^{11})^{\frac{1}{2}} = (40 \times 10^{10})^{\frac{1}{2}} = 40^{\frac{1}{2}} \times 10^{\frac{10}{2}} = \sqrt{40} \times 10^5$$
In this case, you would need to find the root of 40, either by using a square root table or by using logarithms. In Chapter Six, you will learn how to find roots by using logarithms. For now, we will use only numbers with familiar roots.

$$(2.7 \times 10^{-11})^{\frac{1}{3}} = (27 \times 10^{-12})^{\frac{1}{3}} = 27^{\frac{1}{3}} \times 10^{-\frac{12}{3}} = 3 \times 10^{-4}$$

$$(8.1 \times 10^9)^{\frac{1}{4}} = (81 \times 10^8)^{\frac{1}{4}} = (81)^{\frac{1}{4}} \times 10^{\frac{8}{4}} = 3 \times 10^2$$

PROBLEMS. Solve for the following.

(a) $(8 \times 10^{15})^{\frac{1}{3}} =$ _____

(b) $(1.6 \times 10^{-23})^{\frac{1}{2}} =$ _____

(c) $(3.2 \times 10^{41})^{\frac{1}{5}} =$ _____

— — — — — — — — — — — — — — — —

(a) 2×10^5; (b) 4×10^{-12}; (c) 2×10^8

RAISING TO A POWER

(7) To raise to a power in scientific notation, raise the nonexponential part to the power. Then raise the exponential part to the same power. Finally, combine the two answers and convert the final answer to standard form if necessary.

Examples: $(4 \times 10^2)^3 = 4^3 \times 10^{2 \cdot 3} = 64 \times 10^6 = 6.4 \times 10^7$

$(1.2 \times 10^{-8})^2 = (1.2)^2 \times 10^{-8 \cdot 2} = 1.44 \times 10^{-16}$

PROBLEMS.

(a) $(5 \times 10^{-2})^2 =$ _____

(b) $(2 \times 10^3)^4 =$ _____

— — — — — — — — — — — —

(a) $25 \times 10^{-4} = 2.5 \times 10^{-3}$; (b) $16 \times 10^{12} = 1.6 \times 10^{13}$

(8) Before we end this chapter, try the following problems to see if you can put it all together.

Hint: The order of operations is as follows:
(1) Take roots and powers first.
(2) Multiply or divide.
(3) Add or subtract.

PROBLEMS. Solve the following expressions for x.

(a) $x = \dfrac{(2 \times 10^5)(6.0 \times 10^{-13})}{(4 \times 10^{-15})}$ _____

(b) $x = (8.1 \times 10^{-4})(2 \times 10^{-6}) + (3 \times 10^{-11})$ _____

(c) $x = (2 \times 10^{-5})^2 (4 \times 10^{-2})$ _____

(d) $x^5 = 0.4(8 \times 10^{16})$ _____

— — — — — — — — — — — —

(a) $\dfrac{12 \times 10^{-8}}{4 \times 10^{-15}} = 3 \times 10^7$

(b) $16.2 \times 10^{-10} + 3 \times 10^{-11} = 1.62 \times 10^{-9} + .03 \times 10^{-9} = 1.65 \times 10^{-9}$

(c) $(4 \times 10^{-10})(4 \times 10^{-2}) = 16 \times 10^{-12} = 1.6 \times 10^{-11}$

(d) $x^5 = 3.2 \times 10^{16}$
$(x^5)^{\frac{1}{5}} = (3.2 \times 10^{16})^{\frac{1}{5}}$
$x = (32 \times 10^{15})^{\frac{1}{5}}$
$x = 32^{\frac{1}{5}} \times 10^{\frac{15}{5}}$
$x = 2 \times 10^3$

CHAPTER SIX

Logarithms

PRETEST AND OBJECTIVES

This pretest outlines the objectives for Chapter Six. It should help you identify what parts of the chapter, if any, you need to read. Try each of the problems which follow. Then check your answers with the answer key, and follow the directions given there.

Objectives and Pretest   **Your Answer**

When you complete this chapter, you will be able to:

1. Write the logs of simple powers of 10.

    (a) $1.00 =$ _____

    (b) $1000 =$ _____

    (c) $1,000,000 =$ _____

    (d) $1 \times 10^{13} =$ _____

    (e) $0.0001 =$ _____

    (f) $1 \times 10^{-13} =$ _____

2. Use log tables to find the logs of given numbers.

    (a) $999.8 =$ _____

    (b) $27.6 \times 10^{-13} =$ _____

    (c) $0.7831 =$ _____

    (d) $1.234 \times 10^{21} =$ _____

    (e) $1234 =$ _____

    (f) $1.234 =$ _____

3. Using log tables, find the antilogs of given logs.

    (a)   2 + 0.9995 =

    (b)   0 + 0.6990 =

    (c)   −11 + 0.4871 =

    (d)   −7 + 0.4472 =

    (e)   3 + 0.7152 =

    (f)   21 + 0.5551 =

4. Use logs and antilogs in computations.

    (a)   416.9 x 9,760 x 0.00314 =

    (b) $\dfrac{0.00984}{0.00076} =$

    (c) $(0.035)^5 =$

    (d) $(0.0031)^{\frac{1}{2}} =$

5. Find the antilogs of negative logs.

    (a)   −4.8312 =

    (b)   −11.6712 =

    (c)   −7.7777 =

6. Use logs to solve pH problems.

    (a)   What is the pH of pure water?

    (b)   What is the pH of 3.4×10^{-5} M NaOH solution?

    (c)   Find $[H^+]$ of a solution with a pH of 2.8.

Answer Key and Directions

Check your answers with the answer key which follows. If you had difficulty working any of the problems or if you want help with a particular objective, read the indicated portion of Chapter Six.

If you missed any of these problems: **See frames:**

1. (a)   0.0
   (b)   3.0
   (c)   6.0
   (d)   13.0
   (e)   −4.0
   (f)   −13.0   frames 1-3, page 80

2 (a)   2 + 0.9999
   (b)   −12 + 0.4409
   (c)   −1 + 0.8939
   (d)   21 + 0.0913
   (e)   3 + 0.0913
   (f)   0 + 0.0913   frames 4-6, page 82

3. (a) 9.998×10^2
   (b)   5.000
   (c) 3.07×10^{-11}
   (d) 2.80×10^{-7}
   (e) 5.19×10^3
   (f) 3.59×10^{21} frames 7-8, page 88

4. (a) 1.28×10^4
   (b) 1.29×10^1
   (c) 5.25×10^{-8}
   (d) 5.57×10^{-2} frames 9-13, page 91

5. (a) 1.47×10^{-5}
   (b) 2.13×10^{-12}
   (c) 1.67×10^{-8} frame 14, page 100

6. (a)   7
   (b)   4.47
   (c) $1.58 \times 10^{-3}\,M$ frames 15-16, page 101

INTRODUCTION

Logarithms, or logs, are used to handle many complex arithmetic operations quickly and precisely. For example, the following complex problem can be handled in two ways.

$$\frac{1764 \times 0.00392 \times 84.721}{1.987 \times \sqrt{946.3} \times (11.34)^3}$$

(1)   by tedious, long-winded multiplication and division
(2)   by logs

Hopefully, as you read through this chapter, you will become convinced that method (2) is the quickest and most economical alternative.

A log is an exponent. From Chapters Four and Five we recall that exponential arithmetic was a snap. When we take the log of a number, we are converting it to exponential form and thus making our arithmetic easier.

LOGS THAT ARE SIMPLE POWERS OF TEN

① A *log* of a number is the *power* to which 10 must be raised to equal that number. For example, if we are seeking the log of 100, then we are seeking the exponent to which the number 10 must be raised to equal 100.

$100 = 10^2$ (The exponent of 10 that defines the number 100 is 2, and hence, log 100 = 2; that is, the log of 100 is 2.)

To convert any number to a log, we follow a three-step procedure.

Step 1. Rewrite the number in the standard form of scientific notation. (Read Chapter Five if you have forgotten how.)

Step 2. Find the log of each part separately, remembering that scientific notation gives us a two-part quantity (a number part and an exponential part).

$100 = 1 \times 10^2$

$\log (1 \times 10^2) = \log 1 + \log 10^2$ (See Step 3.)

$\log 1 = 0.0$ (Remember, any number $\neq 0$, to the zero power = 1; $10^{0.0} = 1$.)

$\log 10^2 = 2$

Step 3. Combine the logs by adding. Recall that logs are really exponents of 10 for two factors and when you multiply factors, you add exponents.

$1 \times 10^2 = 10^{0.0} \times 10^2 = 10^{2+0.0}$

$\log (1 \times 10^2) = \log 1 + \log 10^2$

$\log (1 \times 10^2) = 0.0 + 2$ or log 100 = 2 + 0.0
(Note: Usually the sum (2 + 0.0) is carried out to give 2.0, but keeping the two parts separate will aid us when we are doing long, consecutive calculations.)

Example: Find the log of .0001.

Step 1. Rewrite .0001 in standard scientific notation.
$.0001 = 1 \times 10^{-4}$

Step 2. Find the log of each part.
$$\log 1 = \log 10^0 = 0$$
$$\log 10^{-4} = -4$$

Step 3. Add the two results.
$$\log (.0001) = \log 1 + \log 10^{-4} = 0 - 4 = -4$$

PROBLEM. Find the log of 1,000,000.

— — — — — — — — — — — — — —

Step 1. $1,000,000 = 1 \times 10^6$
Step 2. $\log (1 \times 10^6) = \log 1 + \log 10^6$
    $\log 1 = 0.0$
    $\log 10^6 = 6$
Step 3. $\log (1 \times 10^6) = 6 + 0.0$ or 6.0

(2)   PROBLEM. Find the log of 0.001.

— — — — — — — — — — — — — —

Step 1. $0.001 = 1 \times 10^{-3}$
Step 2. $\log (1 \times 10^{-3}) = \log 1 + \log 10^{-3}$
    $\log 1 = 0.0$
    $\log 10^{-3} = -3$
Step 3. $\log (1 \times 10^{-3}) = -3 + 0.0$ or -3.0

(3)   PROBLEM. Complete the following table with the scientific notation and the log of each number. The first one has been done as an example.

Number	Scientific Notation	Log
1,000,000	1.0×10^6	$6 + 0.0$
100,000	_____	_____
10,000	_____	_____
1,000	_____	_____
100	_____	_____
10	_____	_____
1	_____	_____
0.1	_____	_____

Number	Scientific Notation	Log
0.01	_____	_____
0.001	_____	_____
0.0001	_____	_____
0.00001	_____	_____
0.000001	_____	_____

– – – – – – – – – – – – – – – –

1.0×10^6	$6 + 0.0$
1.0×10^5	$5 + 0.0$
1.0×10^4	$4 + 0.0$
1.0×10^3	$3 + 0.0$
1.0×10^2	$2 + 0.0$
1.0×10^1	$1 + 0.0$
1.0×10^0	$0 + 0.0$
1.0×10^{-1}	$-1 + 0.0$
1.0×10^{-2}	$-2 + 0.0$
1.0×10^{-3}	$-3 + 0.0$
1.0×10^{-4}	$-4 + 0.0$
1.0×10^{-5}	$-5 + 0.0$
1.0×10^{-6}	$-6 + 0.0$

LOGS THAT ARE FRACTIONAL POWERS OF TEN

(4) Our experience with scientific notation tells us that all positive numbers can be written as the product of a quantity from 1 through 9.99... and 10 raised to a whole number power.

Examples: $300 = 3.0 \times 10^2$
$0.02 = 2.0 \times 10^{-2}$

Finding the log of the exponential part presents no difficulty, as the log of 10 to any power *is* that power. Determining the log of the number part, (e.g., 3.0 and 2.0 above) is more difficult and requires the use of log tables.

Example: Find the log of 300

Step 1. $300 = 3.0 \times 10^2$
Step 2. $\log (3.0 \times 10^2) = \log 3 + \log 10^2$
$\log 3 = ?$
$\log 10^2 = 2$

Now to complete Step 2, we must find a log of a number from 1 through 9.99.... This number will be a decimal fraction (between 0 and

1) between the log 1 = 0.0 and the log 10^1 = 1.0. To do this, look at the sample table which follows. (The exponents of 10 in the table were compiled from more complete log tables. The next section will discuss how to use log tables.)

Number	=	$10^{exponent}$	Exponent	=	Log (number)
1.0	=	$10^{0.0000}$	0.0000	=	log (1)
2.0	=	$10^{0.3010}$	0.3010	=	log (2)
3.0	=	$10^{0.4771}$	0.4771	=	log (3)
4.0	=	$10^{0.6021}$	0.6021	=	log (4)
5.0	=	$10^{0.6990}$	0.6990	=	log (5)
6.0	=	$10^{0.7782}$	0.7782	=	log (6)
7.0	=	$10^{0.8451}$	0.8451	=	log (7)
8.0	=	$10^{0.9031}$	0.9031	=	log (8)
9.0	=	$10^{0.9542}$	0.9542	=	log (9)
10.0	=	$10^{1.0000}$	1.0000	=	log (10)

We note that the number 3.0 = $10^{0.4771}$. With this information we can complete Step 2.

$$\log (3.0 \times 10^2) = \log 10^{0.4771} + \log 10^2$$

$$\log 3.0 = \log 10^{0.4771} = 0.4771$$
$$\log 10^2 = 2$$

Step 3. $\log (3.0 \times 10^2) = 2 + 0.4771$ or 2.4771

Example: Find the log of .009

Step 1. .009 = 9×10^{-3}

Step 2. $\log (9 \times 10^{-3}) = \log 9 + \log 10^{-3}$
$\log 9 = .9542$ (see chart)
$\log 10^{-3} = -3$

Step 3. $\log (.009) = .9542 - 3$, or -2.0458

In other words $10^{-2.0458}$ = .009. Usually, we leave the answer as .9542 − 3 because it makes finding antilogs (which you will read about later) easier. However, it is important that you always remember that a log gives you the exponent that 10 must be raised to for it to equal your original number.

PROBLEMS. Find the logs of the following, using the log table above. Use a separate sheet of paper if needed for your steps of calculation.

(a) 8,000 _____

(b) 0.0008 = _____

(c) 0.006 = _____

(d)   20,000 = _____

— — — — — — — — — — — — — —

(a)   (1) $8,000 = 8 \times 10^3$
      (2) $\log (8 \times 10^3) = \log 8 + \log 10^3$
            $\log 8 = \log 10^{0.9031} = 0.9031$
            $\log 10^3 = 3.0$
      (3) $\log (8 \times 10^3) = 3.0 + 0.9031$ or 3.9031

(b)   (1) $0.0008 = 8 \times 10^{-4}$
      (2) $\log (8 \times 10^{-4}) = \log 8 + \log 10^{-4}$
            $\log 8 = 0.9031$
            $\log 10^{-4} = -4.0$
      (3) $\log (8 \times 10^{-4}) = -4.0 + 0.9031$, or -3.0969

(c)   (1) $0.006 = 6 \times 10^{-3}$
      (2) $\log (6 \times 10^{-3}) = \log 6 + \log 10^{-3}$
            $\log 6 = 0.7782$
            $\log 10^{-3} = -3.0$
      (3) $\log (6 \times 10^{-3}) = -3.0 + 0.7782$ or -2.2218

(d)   (1) $20,000 = 2 \times 10^4$
      (2) $\log (2 \times 10^4) = \log 2 + \log 10^4$
            $\log 2 = 0.3010$
            $\log 10^4 = 4.0$
      (3) $\log (2 \times 10^4) = 4.0 + 0.3010$ or 4.3010

USING LOG TABLES TO FIND LOGS

5   Now we will use more complete log tables.

Example: Find the log of 38,500.

Step 1. $38,500 = 3.85 \times 10^4$
Step 2. $\log (3.85 \times 10^4) = \log 3.85 + \log 10^4$
          $\log 10^4 = 4.0$
          $\log 3.85 = ?$

To find the log of 3.85 we will use the log tables in the Appendix. First, find the first two numbers in the left hand column of the log tables. (Often this column is designated the N column.)

N	0	1	2	3	4	5	6	7	8	9
36
37
38
39
40

Second, move across the row from the number under the column corresponding to the third digit (5).

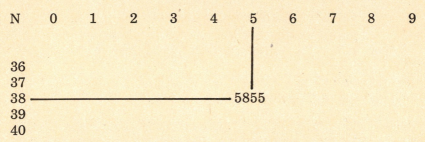

N	0	1	2	3	4	5	6	7	8	9
36
37
38						5855
39
40

The number 5855 constitutes the log of 3.85 and since 3.85 is between 1 and 10, the value of the log of 3.85 will be between 0 and 1, hence:

$$\log 3.85 = 0.5855$$

Now back to the original problem.

Example: Find the log of 38,500.

Step 1. $38,500 = 3.85 \times 10^4$

Step 2. $\log (3.85 \times 10^4) = \log 3.85 + \log 10^4$

   $\log 10^4 = 4.0$

   $\log 3.85 = \log 10^{0.5855} = 0.5855$

Step 3. $\log (3.85 \times 10^4) = 4.0 + 0.5855$ or 4.5855

Again, note that we might choose to keep the number as 4.0 + 0.5855, rather than 4.5855 which is just as correct. We keep these two parts separate because, as we will soon see, it is convenient to manipulate each separately. The whole number before it is combined with the decimal, in this case 4.0, is called the *characteristic* and, as we have seen, may be either positive or negative. The decimal part obtained from log tables, in this case 0.5855, is called the *mantissa* and is *always* positive.

PROBLEMS. Find the logs of the following, using the log tables in the Appendix. Use a separate sheet of paper if needed for your steps of calculation.

(a) 17,900 = _____

(b) 0.000085 = _____

(c) 36,000,000 = _____

(d) 0.00124 = _____

– – – – – – – – – – – – – – – –

(a) (1) $17{,}900 = 1.79 \times 10^4$
   (2) $\log (1.79 \times 10^4) = \log 1.79 + \log 10^4$
       $\log 1.79 = 0.2529$ (from log table)
       $\log 10^4 = 4.0$
   (3) $\log (1.79 \times 10^4) = 4.0 + 0.2529$, or 4.2529

(b) (1) $0.000085 = 8.5 \times 10^{-5}$
   (2) $\log (8.5 \times 10^{-5}) = \log 8.5 + \log 10^{-5}$
       $\log 10^{-5} = -5.0$
       $\log 8.5 = 0.9294$
   (3) $\log (8.5 \times 10^{-5}) = -5 + 0.9294$
       (Here the answer is left in terms of mantissa and character-istic. If we combine the parts, we would get -4.0706, the exponent needed for 10 to become .000085.)

(c) (1) $36{,}000{,}000 = 3.6 \times 10^7$
   (2) $\log (3.6 \times 10^7) = \log 3.6 + \log 10^7$
       $\log 10^7 = 7.0$
       $\log 3.6 = 0.5563$
   (3) $\log (3.6 \times 10^7) = 7.0 + 0.5563$

(d) (1) $0.00124 = 1.24 \times 10^{-3}$
   (2) $\log (1.24 \times 10^{-3}) = \log 1.24 + \log 10^{-3}$
       $\log 10^{-3} = -3.0$
       $\log 1.24 = 0.0934$
   (3) $\log (1.24 \times 10^{-3}) = -3 + 0.0934$, or -2.9066

PROPORTIONAL PARTS TABLE

6   Suppose you have a four-digit number such as 2.563 and you wish to determine the logarithm. To do so, you would use a proportional parts table. Carry out the following steps.

First, find the log of the first three digits (2.56).

log 2.56 = 0.4082

Second, proceed across the row corresponding to the first two digits into the proportional parts column corresponding to the fourth digit (3).

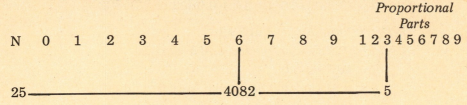

Proportional Parts

| N | 0 | 1 | 2 | 3 | 4 | 5 | 6 | 7 | 8 | 9 | 1 2 3 4 5 6 7 8 9 |

25 —————————————— 4082 ——————————— 5

The proportional part in this case is 0.0005 (the decimal and zeroes are understood).

The third step is to add the log of the first three digits to the proportional part.

log 2.56 = 0.4082

pp = 0.0005
_____
log 2.563 = 0.4087

Example: Find the log of 1.132×10^{-21}

Step 1. $1.132 \times 10^{-21} = 1.132 \times 10^{-21}$
Step 2. $\log (1.132 \times 10^{-21}) = \log 1.132 + \log 10^{-21}$
$\log 10^{-21} = -21.0$
$\log 1.132 = \log 1.13 +$ prop. part from 0.002
$\log 1.132 = 0.0531 + 0.0008 = 0.0539$
Step 3. $\log (1.132 \times 10^{-21}) = -21.0 + 0.0539$

PROBLEMS. What are the logs of the following numbers?

(a)   0.7832 = _____

(b)   956.7 = _____

(c)   0.000003915 = _____

(d)   8.175 = _____

— — — — — — — — — — — — — —

(a)   (1) $0.7832 = 7.832 \times 10^{-1}$
(2) $\log (7.832 \times 10^{-1}) = \log 7.832 + \log 10^{-1}$
$\log 10^{-1} = -1.0$
$\log 7.832 = \log 7.83 +$ prop. part from 0.002
$\log 7.832 = 0.8938 + .0001 = 0.8939$
(3)   log .7832 = −1.0 + 0.8939, or −0.1061

(b) (1) $956.7 = 9.567 \times 10^2$

(2) $\log(9.567 \times 10^2) = \log 9.567 + \log 10^2$

$\log 10^2 = 2.0$

$\log 9.567 = \log 9.56 + \text{prop. part from } 0.007$

$\log 9.567 = 0.9805 + 0.0003 = 0.9808$

(3) $\log(9.567 \times 10^2) = 2.0 + 0.9808, \text{ or } 2.9808$

(c) (1) $0.000003915 = 3.915 \times 10^{-6}$

(2) $\log(3.915 \times 10^{-6}) = \log 3.915 + \log 10^{-6}$

$\log 10^{-6} = -6.0$

$\log 3.915 = \log 3.91 + \text{prop. part from } 0.005$

$\log 3.915 = 0.5922 + 0.0005 = 0.5927$

(3) $\log(3.915 \times 10^{-6}) = -6.0 + 0.5927, \text{ or } -5.4073$

(d) (1) $8.175 = 8.175 \times 10^0$

(2) $\log(8.175 \times 10^0) = \log 8.175 + \log 10^0$

$\log 10^0 = 0$

$\log 8.175 = \log 8.17 + \text{prop. part from } 0.005$

$\log 8.175 = 0.9122 + .0003 = 0.9125$

(3) $\log 8.175 = 0 + 0.9125$

ANTILOGS

7   The next manipulation we must learn before we can do calculations with logs is how to "undo" all our fine work and convert logs back to numbers. This is called *taking the antilog.*

Examples: antilog of $(7.0 + 0.0) = 1 \times 10^7$

antilog of $(2.0 + 0.0) = 1 \times 10^2$

antilog of $(-7.0 + 0.0) = 1 \times 10^{-7}$

antilog of $(-20.0 + 0.0) = 1 \times 10^{-20}$

This makes sense. Since the exponent of 10 is the log, the antilog of a characteristic is 10 raised to a power equal to the characteristic. We know that the log of 1.000 is 0.000, so the antilog of 0.0000 must be 1.

But how do we find antilogs of numbers with mantissas greater than 0.0000? This is done by finding the log in the log table. The row in which the log resides gives the first two digits, and the column in which you find the number gives the third digit. For example, to find the antilog of 0.8401 proceed as follows:

First, find 8401 in the log table (see the sample table).

The first two digits of the antilog are in the N column of the *row* in which you find the 8401.

N	0	1	2	3	4	5	6	7	8	9
68										
69			8401							
70										

The third digit of the antilog is the heading of the *column* in which you find 8401.

N	0	1	2	3	4	5	6	7	8	9
68										
69			8401							
70										

Therefore, the antilog of 0.8401 = 6.92

We know, of course, that the number is 6.92, not 692 or 0.692 or some other combination, since we applied scientific notation initially to get into logs. To be certain we have the correct antilog, quickly reverse the process for a check and take the log of 6.92. We find 0.8401 is indeed the log, so we carried out the antilog manipulation correctly.

Examples: Find the antilog of 7.0 + 0.5490 (or 7.5490).
antilog (7.0 + 0.5490) = antilog 7 x antilog 0.5490
antilog 7 = 10^7
antilog 0.5490 = 3.54 (from log tables)
antilog (7.0 + 0.5490) = 3.54 x 10^7

Find the antilog of −5.0 + 0.5490.
antilog (−5.0 + 0.5490) = antilog −5.0 x antilog 0.5490
antilog −5.0 = 10^{-5}
antilog 0.5490 = 3.54
antilog (−5.0 + 0.5490) = 3.54 x 10^{-5}

Now you see why we usually write the log as −5.0 + 0.5490 rather than −4.4510. It is easier to find the antilog when the mantissa and characteristic are intact and haven't been combined.

Find the antilog of 0 + 0.9304.
antilog (0 + 0.9304) = antilog 0 x antilog 0.9304
antilog 0 = 10^0
antilog 0.9304 = 8.52 (from log tables)
antilog (0 + 0.9304) = 8.52 x 10^0 = 8.52

PROBLEMS. Find the antilogs (abbreviated al.) of the following. Use a separate sheet of paper to write your steps.

(a) $0 + 0.0043 =$ _____

(b) $5 + 0.4871 =$ _____

(c) $-7.0 + 0.9991 =$ _____

— — — — — — — — — — — — — —

(a)   al. $(0.0 + 0.0043) =$ al. $(0.0) \times$ al. (0.0043)
al. $(0.0) = 1.0$
al. $(0.0043) = 1.01$
al. $(0.0) \times$ al. $(0.0043) = 1.0 \times 1.01 = 1.01$

(b)   al. $(5.0 + 0.4871) =$ al. $(5.0) \times$ al. (0.4871)
al. $(5.0) = 10^5$
al. $(0.4871) = 3.07$
al. $(5.0 + 0.4871) = 3.07 \times 10^5$

(c)   al. $(-7.0 + 0.9991) =$ al. $(-7.0) \times$ al. (0.9991)
al. $(-7.0) = 10^{-7}$
al. $(0.9991) = 9.98$
al. $(-7.0 + 0.9991) = 9.98 \times 10^{-7}$

ANTILOGS AND PROPORTIONAL PARTS

8   If we wish to find the antilog of 4835 and we look in the table we will not find 4835 but we will find 4829 and 4843. What to do? We can use the proportional parts table in reverse but that is a lot of trouble.

If we should require more than three significant figures, it is easier to look in a five- or seven-place log table than to work proportional parts in reverse. Five-place log tables are available, for example, in the *Handbook for Chemistry and Physics* and seven-place tables are available in most libraries. These larger tables work just like the smaller ones except that the N column contains three or four digits instead of two.

Most introductory science classes only require three significant figures, so the best thing to do is to take that number for the antilog whose log is closest to the number. (If they are equal take the lower.)

For example, the antilog of 4829 is 3.04 and the antilog of 4843 is 3.05. Now 4835 is 6 from 4829 and 8 from 4843, so the antilog of 4835 will be closer to 3.04 than to 3.05.

Examples: What is the antilog of 3.0 + 0.7150?
al. (3.0 + 0.7150) = al. (3.0) x al. (0.7150)
al. (3.0) = 10^3
al. (0.7150) = 5.19 (In the log table we find 0.7143.
and 0.7152, and 0.7150 is closer to 0.7152.)
al. (3.0 + 0.7150) = 5.19 x 10^3

Find the antilog of (−7.0 + 0.4471).
al. (−7.0 + 0.4471) = al. (−7.0) x al. (0.4471)
al. (−7.0) = 10^{-7}
al. (0.4471) = 2.8 (closest to 0.4472)
al. (−7.0 + 0.4471) = 2.8 x 10^{-7}

PROBLEMS. Find the antilog of the following. Use a separate sheet of paper for your steps.

(a)   5.0 + 0.8020 =   _____

(b)   −7.0 + 0.8925 =   _____

— — — — — — — — — — — — —

(a)   al. (5.0 + 0.8020) = al. (5.0 x al. (0.8020)
al. (5.0) = 10^5
al. (0.8020) = 6.34 (closest to 0.8021)
al. (5.0 + 0.8020) = 6.34 x 10^5

(b)   al. (−7.0 + 0.8925) = al. (−7.0) x al. (0.8925)
al. (−7.0) = 10^{-7}
al. (0.8925) = 7.81 (closest to 0.8927)
al. (−7.0 + 0.8925) = 7.81 x 10^{-7}

THE USE OF LOGS IN PROBLEMS

⑨   We know that logs are exponents of 10, and previously (Chapter Four) we learned how to perform operations on like quantities raised to different powers. To apply our knowledge of logs to problem solving, we should remember the following rules.

Rule 1. To multiply numbers, take the log of each, add them, and find the antilog.

Example: 12 x 11 =

Number	Log
1.2×10^1	1.0 + 0.0792
1.1×10^1	1.0 + 0.0414
	2.0 + 0.1206

al. $(2.0 + 0.1206) = $ al. (2.0) x al. (0.1206)
$$= 10^2 \times 1.32$$
and $12 \times 11 = 1.32 \times 10^2$

(Remember our rules governing operations on exponents?)

$1.2 \times 10^1 = 10^{1.0} \times 10^{0.0792}$
$\underline{1.1 \times 10^1 = 10^{1.0} \times 10^{0.0414}}$

$12 \times 11 = 10^{2.0} \times 10^{0.1260}$
$12 \times 11 = 10^2 \times 1.32 = 132$

Example: Carry out the following multiplications using logs.

$0.000712 \times (271 \times 10^{-23}) \times (9.32 \times 10^{-25}) \times 4.12$

Number	Log
7.12×10^{-4}	$- 4 + 0.8525$
2.71×10^{-21}	$-21 + 0.4330$
9.32×10^{-25}	$-25 + 0.9694$
4.12×10^0	$\underline{- 0 + 0.6149}$
	$-50 + 2.8698 = -48 + 0.8698$

al. $(-48 + 0.8698) = $ al. (-48) x al. $(0.8698) = 7.41 \times 10^{-48}$

Note how the sum of the mantissas gave us additional values to be added to the characteristic.

PROBLEMS. Carry out the following multiplications using logs. Use a separate sheet of paper for your calculations.

(a) $416.9 \times 9,760 \times 0.00314 = $ _____

(b) $487 \times 2.45 \times 0.0387 = $ _____

— — — — — — — — — — — — —

(a)
Number	Log
4.169×10^2	$2.0 + 0.6200$
9.760×10^3	$3.0 + 0.9894$
3.14×10^{-3}	$\underline{-3.0 + 0.4969}$
	$2.0 + 2.1063 = 4.0 + 0.1063$

al. $(4.0 + 0.1063) = 1.28 \times 10^4$ (Note: 0.1072 is the closest log.)

(b)
Number	Log
4.87×10^2	$2.0 + 0.6875$
2.45×10^0	$0.0 + 0.3892$
3.87×10^{-2}	$\underline{-2.0 + 0.5877}$
	$0.0 + 1.6644 = 1.0 + 0.6644$

al. $(1.0 + 0.6644) = 4.62 \times 10^1$ (Note: 0.6646 is closest log.)

(10) Our second rule of problem solving is as follows.
Rule 2. To divide numbers, take the log of each and subtract the denominator from the numerator and take the antilog of the result.

> Example: Divide 132 by 11
>
> | Number | Log |
> |---|---|
> | 1.32×10^2 | $2 + 0.1206$ |
> | $1.1 \ \ \times 10^1$ | $-(1 + 0.0414)$ |
> | | $1 + 0.0792$ |
>
> al. $(1 + 0.0792) = 1.2 \times 10^1 = 12$
>
> This is the same as: $\dfrac{132}{11} = \dfrac{10^{0.1206} \times 10^2}{10^{0.0414} \times 10^1}$
>
> $$= 10^{0.0792} \times 10^1$$

> Example: $\dfrac{1.764 \times 10^{-10}}{3.89 \times 10^{-6}}$
>
> | Number | Log |
> |---|---|
> | 1.764×10^{-10} | $-10 + 0.2465$ |
> | $3.89 \ \ \times 10^{-6}$ | $-(-6 + 0.5899)$ |
>
> > At this point, we see that if we subtract 0.5899 from 0.2465 we will obtain a negative mantissa and log tables give only positive mantissas. To avoid this difficulty it is permissible to add 1 to the mantissa from the numerator and subtract 1 from the characteristic.
>
> | | |
> |---|---|
> | -11 | $+ 1.2465$ |
> | $-(-6$ | $+ 0.5899)$ |
> | $- 5$ | $+ 0.6566$ |
>
> al. $(-5 + 0.6566) = 4.53 \times 10^{-5}$
>
> > (Note: 0.6566 is halfway between 0.6561 and 0.6571 so we take the lower log, 0.6561, which gives us 4.53.)

PROBLEMS. Solve the following using logs.

(a) $\dfrac{0.00984}{0.00076} = $ _____

(b) $\dfrac{1.983 \times 1842}{0.0071 \times 847} = $ _____

_ _ _ _ _ _ _ _ _ _ _ _ _ _ _

(a) *Number*   *Log*
9.84 x 10^{-3} $-3 + 0.9930$
7.60 x 10^{-4} $-(-4 + 0.8808)$
————————————
1 + 0.1122

al. (1 + 0.1122) = 1.29 x 10^1 or 12.9

(Note: 0.1122 is closest to 0.1106 in the log table.)

(b) for 1.983 x 1842:   *Number*   *Log*
1.983 x 10^0 0 + 0.2974
1.842 x 10^3 3 + 0.2653
————————
3 + 0.5627

for 0.0071 x 847:   *Number*   *Log*
7.1 x 10^{-3} $-3 + 0.8513$
8.47 x 10^2 $2 + 0.9279$
————————————
$-1 + 1.7792$ or 0 + 0.7792

3 + 0.5627
$-(0 + 0.7792)$
————————

To avoid negative mantissa we write:

2 + 1.5627
$-(0 + 0.7792)$
————————
2 + 0.7835

al. (2 + 0.7835) = 6.07 x 10^2

(11) We continue with our problem solving rules.

Rule 3. To raise any number to a higher power, take the log of the number, multiply by the power, and then find the antilog. (Remember, $(3^2)^3 = 3^6$. A log is an exponent, so if we raise it to a power, we multiply by the power.)

Example: Find by logs: 3^3
Number   *Log*
$(3.0 \times 10^0)^3$ 0 + 0.4771
x 3
————————
0 + 1.4313 = 1 + 0.4313

al. (1 + 0.4313) = 2.7 x 10^1 = 27

PROBLEMS. Solve using logs.

(a) $(0.035)^5 =$ _____

(b) $(21.14)^3 =$ _____

(c) $\dfrac{113.3 \times (599 \times 10^3)^6}{0.0954} =$ _____

(Hint: According to order of operations, you should multiply before you add or subtract.)

_ _ _ _ _ _ _ _ _ _ _ _ _ _ _

(a) Number   Log

$(3.5 \times 10^{-2})^5$ $-2 + 0.5441$

                           $\underline{\times\ 5}$

         $-10 + 2.7205 = -8 + 0.7205$

al. $(-8 + 0.7205) = 5.25 \times 10^{-8}$

     (0.7205 is closest to 0.7202 in the log table.)

(b) Number   Log

$(2.114 \times 10^1)^3$ $1 + 0.3251$

                          $\underline{\times\ 3}$

          $3 + 0.9753$

al. $(3 + 0.9753) = 9.45 \times 10^3$

     (0.9753 is closest to 0.9754 in the log table.)

(c) Number   Log

$(5.99 \times 10^5)^6$ $5 + 0.7774$

                          $\underline{\times\ 6}$

          $30 + 4.6644$ $\Big\}$ *add*

1.133×10^2 $\underline{2 + 0.0542}$

          $32 + 4.7186$ $\Big\}$ *subtract*

9.54×10^{-2} $\underline{-(-2 + 0.9795)}$

          $34 + 3.7391 = 37 + 0.7391$

al. $(37 + 0.7391) = 5.48 \times 10^{37}$

     (0.7391 is closest to 0.7388 in the log table.)

(12) Following is the fourth rule of problem solving.

Rule 4. To find the root of a number, take the log of the quantity, divide by the root, and find the antilog of the quotient. (Remember, logs are exponents. When we take the root of an exponent we multiply by $\dfrac{1}{\text{root}}$ which is the same as dividing by the root.)

     Examples: Find $\sqrt{144}$ by logs.

               Number   Log

              $(1.44 \times 10^2)^{\frac{1}{2}}$ $\dfrac{2 + 0.1584}{2}$

                             $1 + 0.0792$

         al. $(1 + 0.0792) = 1.2 \times 10^1 = 12$

Find $(0.000341)^{\frac{1}{4}}$ by logs.

Number	Log
$(3.41 \times 10^{-4})^{\frac{1}{4}}$	$\dfrac{-4 + 0.5328}{4}$
	$-1 + 0.1332$

al. $(-1 + 0.1332) = 1.36 \times 10^{-1} = 0.136$

(0.1332 is closest to 0.1335 in the log table.)

PROBLEMS. Find the following using logs.

(a) $(256)^{\frac{1}{4}} =$ _____

(b) $(1.85)^{\frac{1}{2}} =$ _____

— — — — — — — — — — — — — —

(a)
Number	Log
$(2.56 \times 10^2)^{\frac{1}{4}}$	$\dfrac{2 + 0.4082}{4}$

al. $(0.6021) = 4$

(b)
Number	Log
$(1.85 \times 10^0)^{\frac{1}{2}}$	$\dfrac{0 + 0.2672}{2}$

al. $(0.1336) = 1.36$

(0.1336 is closest to 0.1335 in the log table.)

(13)   Sometimes the problems are more difficult.

Example: Find $(5.5 \times 10^4)^{\frac{1}{3}}$ by logs.

Number	Log
$(5.5 \times 10^4)^{\frac{1}{3}}$	$\dfrac{4 + 0.7404}{3}$

al. $(1 + .5801) = 3.80 \times 10^1 = 38$

(0.5801 is closest to 0.5798 in the log table.)

Here we had a remainder of 1 from dividing 3 into the characteristic 4, and we carried it to the mantissa and continued with our division on 1.7404. Had the characteristic been negative, we would have been in a dilemma. The remainder would have been negative while our mantissa was positive. To avoid complications of this type, we use a "trick" whenever we encounter a negative characteristic that is not evenly divisible by the root.

Trick or Subrule 4. When taking the root of a number that gives a log with a *negative characteristic* that is not evenly divided by the root we make the characteristic more negative in 1-unit increments until it is evenly divided by the root. At the same time we increase the mantissa by a +1 unit for each unit change in the characteristic. (Note that the net effect is 0 because every time you add −1 to the characteristic, you add +1 to the mantissa.)

Examples: Find $(0.0031)^{\frac{1}{2}}$ by logs.

Number	Log
$(3.1 \times 10^{-3})^{\frac{1}{2}}$	$\dfrac{-3 + 0.4914}{2}$

$$\dfrac{-4 + 1.4914}{2} \quad \text{(Decrease character-}$$

istic by −1, increase mantissa by +1.)

al. $(-2 + .7457) = 5.57 \times 10^{-2}$

(0.7457 is closest to 0.7459 in the log table.)

Find $(4.2 \times 10^{-16})^{\frac{1}{3}}$

Number	Log
$(4.2 \times 10^{-16})^{\frac{1}{3}}$	$\dfrac{-16 + 0.6232}{3}$

$$\dfrac{-17 + 1.6232}{3} \quad \text{(−17 is not evenly}$$

divisible, so try another increment.)

$$\dfrac{-18 + 2.6232}{3}$$

al. $(-6 + .8744) = 7.49 \times 10^{-6}$

(0.8744 is closest to 0.8745 in the log table.)

To keep tabulation clear, we work the numerator and denominator separately. Also we follow the order of operations, doing powers and roots (multiplication and division in logarithm computation) first.

Example: Find $\dfrac{(5.2)^3 (.015)}{(82)^2 (1.45)^{\frac{1}{2}}}$

Numerator

Number	Log
$(5.2 \times 10^0)^3$	$0 + 0.7160$
	$\underline{\qquad\qquad \times 3}$
	$0 + 2.1480$ $\quad\big\}$ add
(1.5×10^{-2})	$\underline{-2 + 0.1761}$
	$-2 + 2.3241 \quad = 0 + 0.3241$

Denominator

Number	Log
$(8.2 \times 10^1)^2$	$1 + 0.9138$
	$\times 2$
	$\overline{2 + 1.8276}$
$(1.45 \times 10^0)^{\frac{1}{2}}$	$\dfrac{0 + 0.1614}{2}$
	$0 + 0.0807$

add

$\overline{2 + 1.9083} = 3 + 0.9083$

$$\frac{\text{Numerator}}{\text{Denominator}} = \frac{0 + 0.3241}{-(3 + 0.9083)} = \frac{-1 + 1.3241}{-3 - 0.9083}$$

$$\overline{-4 + 0.4158}$$

al. $(-4 + 0.4158) = 2.60 \times 10^{-4}$ (0.4158 is midway between 0.4150 and 0.4166 in the log table so we take the lowest and obtain 2.60.)

Examples:

Characteristic not evenly divisible

$\dfrac{-3 + 0.0514}{2}$ $\xrightarrow{-1+1}$

$\dfrac{-4 + 0.5163}{3}$ $\xrightarrow{-2+2}$

$\dfrac{-2 + 0.9872}{7}$ $\xrightarrow{-5+5}$

Characteristic evenly divisible

$\dfrac{-4 + 1.0514}{2}$

$\dfrac{-6 + 2.5163}{3}$

$\dfrac{-7 + 5.9872}{7}$

PROBLEMS. Use logs to solve the following.

(a) $(2.4 \times 10^2)^{\frac{1}{4}} =$ _____

(b) $(5.1 \times 10^3)^{\frac{2}{3}} =$ _____

(c) $(72.5 \times 10^{-8})^{\frac{1}{6}} =$ _____

(d) $\dfrac{(0.031)(813 \times 10^3)^4}{(0.021)^{\frac{1}{4}}(718)^5} =$ _____

- - - - - - - - - - - - - - - -

(a)   *Number*

$(2.4 \times 10^2)^{\frac{1}{4}}$

Log

$\dfrac{2 + 0.3802}{4}$

al. $(0 + .5951) = 3.94$

(b)   *Number*   *Log*
      (5.1×10^3) 3 + 0.7076
                                     $\times\ 2$
                                _____
                                6 + 1.4152
                                _____
                                     3

al. $(2 + .4717) = 2.96 \times 10^2$

(c)   *Number*   *Log*
      (7.25×10^{-7}) −7 + 0.8603
                                _____
                                     6

                                  $(-5+5)$
                                −12 + 5.8603
                                _____
                                     6

al. $(-2 + 0.9767) = 9.48 \times 10^{-2}$

(d)   Numerator
      Number   *Log*
      $(8.13 \times 10^5)^4$ 5 + 0.9101
                                     $\times\ 4$
                                _____
                                20 + 3.6404 $\Big\}$ *add*
      (3.1×10^{-2}) −2 + 0.4914
                                _____
                                18 + 4.1318 = 22 + 0.1318

               Denominator
      Number   *Log*
      $(2.1 \times 10^{-2})^{\frac{1}{4}}$ −2 + 0.3222
                                _____
                                     4

                                  $(-2+2)$
                                −4 + 2.3222
                                _____
                                     4

                                −1 + 0.5805
      $(7.18 \times 10^2)^5$ 2 + 0.8561 $\Big\rangle$ add
                                     $\times\ 5$
                                _____
                                10 + 4.2805
                                _____
                                13 + 0.8610

        Numerator   =   22 + 0.1318
      −Denominator   −(13 + 0.8610)
      _____   _____

                           21 + 1.1318
                          −(13 + 0.8610)
                          _____
                            8 + 0.2708

al. $(8 + 0.2708) = 1.87 \times 10^8$

ANTILOGS OF NEGATIVE LOGARITHMS

(14)   At some point in your science career, you will encounter *negative logarithms.* These can cause a little difficulty unless you remember that the mantissa must always be positive.

Suppose, for example, that we are required to take the antilog of -4.8312. Do not confuse this with $-4 + 0.8312$, which is the normal form and has an antilog of 6.78×10^{-4}. We should note that -4.8312 is a negative log and that somehow we must "play" with this number and get it in a form with a positive mantissa so that we can find its antilog.

To do this we go to the next more negative characteristic (in this case, -5.0000) and ask, "What number must we add to -5.0000 so that it is equal to -4.8312?" The answer is 0.1688, so our new number, complete with positive mantissa, is $-5 + 0.1688$.

$$-5 + 0.1688 = -4.8312$$

We can take the antilog of $-5 + 0.1688$, which is in the correct form of characteristic (positive or negative) and mantissa (always positive).

The antilog of $-5 + 0.1688$ is 1.47×10^{-5}. (Note that this is considerably different from the antilog of $-4 + 0.8312$.)

Example: Find the antilog of -11.6712.

Step 1.   Our characteristic will be more negative by 1; i.e., -12.

Step 2.   Our mantissa is obtained by subtracting 0.6712 from 1.0000.

$$
\begin{array}{r}
-\ 1 + 1.0000 \\
-11 - 0.6712 \\
\hline
-12 + 0.3288
\end{array}
$$

al. $(-12 + 0.3288) = 2.13 \times 10^{-12}$

(Note: $-1+1$ has a net effect of 0 so the logarithm is not changed. Also, note that $-11.6712 = -11 - 0.6712$. The decimal portion is negative also.)

Example: Find the antilog of -6.3214.

$$
\begin{array}{r}
-1 + 1.0000 \\
-6 - 0.3214 \\
\hline
-7 + 0.6786
\end{array}
$$

al. $(-7 + 0.6786) = 4.77 \times 10^{-7}$

PROBLEMS. What are the antilogs of the following?

(a) $-0.0004 = $ _____

(b) $-7.7777 = $ _____

(c) $-24.9800 =$ _____

(d) $-1000.1000 =$ _____

$----------------$

(a) $-1 + 1.0000$
  $\underline{-0 - 0.0004}$
  $-1 + 0.9996$
  al. $(-1 + 0.9996) = 9.99 \times 10^{-1}$

(b) $-1 + 1.0000$
  $\underline{-7 - 0.7777}$
  $-8 + 0.2223$
  al. $(-8 + 0.2223) = 1.67 \times 10^{-8}$

(c) $-\ 1 + 1.0000$
  $\underline{-24 - 0.9800}$
  $-25 + 0.0200$
  al. $(-25 + 0.0200) = 1.05 \times 10^{-25}$

(d) $-\ \ \ \ 1 + 1.0000$
  $\underline{-1000 - 0.1000}$
  $-1001 + 0.9000$
  al. $(-1001 + 0.9000) = 7.94 \times 10^{-1001}$

"p" FUNCTIONS

(15) This section will not be of any value until you have been introduced to the concept of pH in your course. You may want to skip this section unless you are taking chemistry or closely related courses such as biochemistry or microbiology.

The "p" functions are defined as the *negative log* of some quantity. The term pH is defined as the negative log of the hydrogen ion concentration, as explained below.

pH $= -\log[H^+]$. The brackets, [ ], mean concentration in moles per liter. Whatever appears in the brackets defines what chemical species, in this case the hydrogen ion, H^+. (Sometimes H^+ concentration is written as the hydronium ion concentration, $[H_3O^+]$; these should be considered equivalent terms.)

Example: The hydrogen ion concentration of pure water is $1 \times 10^{-7} M$. (*M* means moles per liter). What is the pH of water?

pH $= -\log[H^+]$ (By definition.)
pH $= -\log(1 \times 10^{-7})$ (Substitute numbers for $[H^+]$.)

$pH = -(\log 1 + \log 10^{-7})$
$pH = -(0.0 + (-7))$ (Take logs.)
$pH = 7$ (Account for minus sign.)

Example: Find the pH of 0.007 M HCl solution (i.e., find the pH when $[H^+] = 7.0 \times 10^{-3}$ M.

$pH = -\log [H^+]$ (By definition.)
$pH = -\log (7 \times 10^{-3})$ (Substitute numbers.)
$pH = -(\log 7 + \log 10^{-3})$
$pH = -(0.8451 + (-3))$ (Take logs.)
$pH = 3 - 0.8451$ (Account for minus sign.)
$pH = 2.2$ (Round off.)

Example: Find the pOH of $3.4 \times 10^{-5} M$ NaOH solution. That is a solution that has $[OH^-] = 3.4 \times 10^{-5}$ M.

$pOH = -\log [OH^-]$ (By definition.)
$pOH = -\log (3.4 \times 10^{-5})$ (Substitute numbers.)
$pOH = -(\log 3.4 + \log 10^{-5})$
$pOH = -(0.5315 + (-5))$ (Take logs.)
$pOH = 5 - 0.5315$ (Account for minus sign.)
$pOH = 4.47$ (Round off.)

PROBLEMS. What is the pH of each of the following solutions? Use a separate sheet of paper for your calculations.

(a) $[H^+] = 1 \times 10^{-4}$ M pH = _____

(b) $[H^+] = 1 \times 10^{-2}$ M pH = _____

(c) $[H^+] = 0.0025$ M pH = _____

(d) $[H^+] = 7.82 \times 10^{-6}$ M pH = _____

– – – – – – – – – – – – – – – – –

(a) $pH = -\log[H^+]$
$pH = -\log (1 \times 10^{-4})$
$pH = -(\log 1 + \log 10^{-4})$
$pH = -(0.0 + (-4))$
$pH = 4.0$

(b) $pH = -\log (1 \times 10^{-2})$
$pH = -(\log 1 + \log 10^{-2})$
$pH = -(0.0 + (-2))$
$pH = 2.0$

(c) pH = $-\log (2.5 \times 10^{-3})$
pH = $-(\log 2.5 + \log 10^{-3})$
pH = $-(0.3979 + (-3))$
pH = 2.6
(Note: Often pH is rounded to the first decimal.)

(d) pH = $-\log (7.82 \times 10^{-6})$
pH = $-(\log 7.82 + \log 10^{-6})$
pH = $-(0.8932 + (-6))$
pH = $6 - 0.8932$
pH = 5.11

(16) To find the hydrogen ion concentration from pH, we must find the antilog of a negative log. So frame 14 will come in handy, now.

Example: Find $[H^+]$ of a solution with a pH of 2.8.
pH = $-\log [H^+]$
$2.8 = -\log [H^+]$
$-2.8 = \log [H^+]$
$\underline{-2 - 0.8000}$ (Rewrite into conventional log form ac-
$(-1+1)$ cording to frame 14.)
$-3 + 0.20000 = \log [H^+]$
al. $(-3 + 0.2000) =$ al. $(\log [H^+])$
$1.58 \times 10^{-3} M = [H^+]$

PROBLEMS. Find the $[H^+]$ of each of the following solutions.

(a) pH = 5.12 _____

(b) pH = 13.5 _____

(c) pH = 1.1 _____

– – – – – – – – – – – – – – –

(a) pH = $-\log [H^+]$
$5.12 = -\log [H^+]$
$-5.12 = \log [H^+]$
$-5 - 0.1200$
$(-1+1)$
$-6 + 0.8800 = \log [H^+]$
al. $(-6 + 0.8800) =$ al. $(\log [H^+])$
$10^{-6} \times 7.59 = [H^+]$
$[H^+] = 7.59 \times 10^{-6} M$

(b) $pH = -\log [H^+]$
$13.5 = -\log [H^+]$
$-13.5 = \log [H^+]$
$-13 - 0.5000$
$(-1+1)$
$-14 + 0.5000 = \log [H^+]$
al. $(-14 + 0.5000) = $ al. $(\log [H^+])$
$10^{-14} \times 3.16 = [H^+]$
$[H^+] = 3.16 \times 10^{-14} \ M$

(c) $pH = -\log [H^+]$
$1.1 = -\log [H^+]$
$-1.1 = \log [H^+]$
$-1 - 0.1000$
$(-1+1)$
$-2 + 0.9000 = \log [H^+]$
al. $(-2 + 0.9000) = $ al. $(\log [H^+])$
$10^{-2} \times 7.94 = [H^+]$
$[H^+] = 7.94 \times 10^{-2} \ M$

CHAPTER SEVEN

Problem Solving and Dimensional Analysis

This chapter contains no pretest. Most of the concepts are introduced in the first few weeks of most science courses. Some of the concepts (e.g., heat change and heat transfer) are specifically for chemistry students.

The underlying theme of this entire chapter is to demonstrate that problems can be solved by a variety of methods (e.g., dimensional analysis, algebra, rearranging formulas). However, proper reasoning is necessary and that is achieved by practice. Therefore, this chapter is full of practice problems, including additional practice problems at the end.

① 1 Difficulties in solving problems arise from procedural errors in three general areas:

1. Correct identification of the "given" and the "unknown".
2. The method used to translate the "given" into a value for the "unknown".
3. The calculation steps (adding, subtracting, etc.).

The first difficulty can be overcome easily by assuming a disciplined approach and making note of the "given" and "unknown." Until you become completely confident in problem solving, it is best to make the headings below.

Unknown = _____

Given = _____

While reading the problem, you can then fill in the blanks.

Example: Radioactive carbon-14 has a half life of 5,700 years. What fraction of carbon-14 will remain in a fossil after 57,000 years?

Unknown: Fraction of carbon-14 left
Given: Half life = 5,700 years; fossil = 57,000 years.

PROBLEMS. Identify the "Given" and the "Unknown" in the

following questions.

(a) If there are 24 bottles in a case of beer, and one case costs $5.25, how much does one bottle cost?

_____

(b) A 5 liter flask contains 1 mole of a gas at 300°C. What is the pressure of the gas?

_____

_ _ _ _ _ _ _ _ _ _ _ _ _ _ _ _ _

(a) Unknown: Cost of one bottle
     Given: 24 bottles per case; $5.25 per case
(b) Unknown: Pressure
     Given: 5 liters; 300°C; and one mole

This last question may be meaningless to you — do not worry about it, as it was only included to demonstrate how to organize the content of a question. Later on, if you study chemistry, you will learn what the terms *liter*, *mole*, and *pressure* mean.

(2) It is imperative that we exercise care in identifying the "given" since extraneous values and variables will be referred to in some problems. Only our knowledge of the subject will prevent us from trying to include meaningless values in our solutions.

    Example: What is the speed of an automobile that travels 120 miles in two hours during the time of year when the earth is rotating at a rate of 0.001% less than 24 revolutions per day?
        Unknown: Speed
        Given: 120 miles; 2 hours; and rate of rotation

In this problem the rate of rotation of earth is extraneous — it is of no importance to the value of the speed. Generally, the excess information will not be so obvious and you will have to apply your knowledge of the subject to decide what information is relevant and what is not.

PROBLEM. A man spends half of his weekly paycheck on Friday, and spends half of the balance each day. By the next Thursday, he has $2 left. What was his original Friday paycheck if he bought a used car for $1500?

_____

_ _ _ _ _ _ _ _ _ _ _ _ _ _ _ _ _

Unknown: Original paycheck
Given: Spent one-half of money each day; $2 left on Thursday; bought $1500 car (extraneous)

USE OF EQUATIONS

(3)   After defining the elements of a problem, you will generally have
two choices for a method of translating the given and the unknown
into an answer. (Remember, this area was the second potential source
of error described on page 105.) The first is to memorize equations,
correlate the appropriate equation with a question, and then rearrange
it so that the unknown is alone on one side and everything else is on
the other. All of this is followed by substituting values of the "known"
into the equation and then calculating.

Example: Using the equation, distance = speed x time, what is the
speed of an automobile that travels 40 miles in two
hours?

Unknown: Speed
Given: 40 miles (distance); 2 hours (time)
Solution: (1) Supply equation.
                 distance = speed x time
            (2) Rearrange so that the unknown is on
                 one side and everything else is on the
                 other.

$$\frac{distance}{time} = \frac{speed \times \cancel{time}}{\cancel{time}} , \text{ or}$$

$$\frac{distance}{time} = speed$$

            (3) Substitute values for distance and
                 time and then carry out the division.

$$\frac{40 \text{ miles}}{2 \text{ hours}} = speed, \text{ or } \frac{20 \text{ miles}}{hour}$$

$$= speed$$

Notice that the units "mile" and "hour" are included in the answer.
This is essential, for when we report numerical values for a physical
measurement it is meaningless to say "20" or "5,000,000" without
saying 20 zongs or 5 million bleeps, or whatever units we are discus-
sing. As a matter of practice then, you must take great care to include
all the units throughout your calculations. This is a particularly important point if you should use the second method of translating a question into an answer, which is dimensional analysis.

PROBLEM. Deviation IQ is figured by multiplying the number of
standard deviations by 15 and adding the result to 100. How many
standard deviations are required for an IQ score of 125? (Write your
calculation steps on a separate sheet of paper.)

Unknown: Number of standard deviations
Given: IQ = 125
Solution: (1) Equation: IQ = 15(S.D.) + 100

(2) Rearrange equation. $\dfrac{IQ - 100}{15}$ = S.D.

(3) Substitute values.

$$\dfrac{125 - 100}{15} = S.D.$$

$$\dfrac{25}{15} = S.D.$$

S.D. = $+\dfrac{5}{3}$ standard deviations

DIMENSIONAL ANALYSIS

④   *Dimensional analysis* (also called *units analysis*) is a process for converting the dimensions (units) of the given into those of the unknown. To do this, we multiply the given by a ratio that will result in the units required for the answer. We set up the multiplication so that the "unwanted" units (units not called for in the answer) will be divided out by one of the components of the ratio by which we are multiplying.

For example, we know that there are 12 inches in one foot and that one foot is made up of 12 inches, so we have the identity: 12 inches = 1 foot. If we divide both sides of the identity by 1 foot we have:

$$\frac{12 \text{ inches}}{1 \text{ foot}} = \frac{1 \text{ foot}}{1 \text{ foot}} = 1$$

Because 12 inches is equivalent to 1 foot, the expression $\dfrac{12 \text{ inches}}{1 \text{ foot}}$ is equal to 1.

Since we know that multiplying anything by 1 does not change its value, this expression can be used to simplify units.

Example: How many inches are in 3.5 feet?

Unknown: Inches
Given: 3.5 feet
Method: Dimensional analysis

12 inches = 1 foot, so $\dfrac{12 \text{ inches}}{1 \text{ foot}} = 1$

(From now on, add this "Method" step to your problem-solving procedure. Here we wanted to choose a factor that allows us to divide out feet and keep inches in the numerator.)

Solution: Multiply 3.5 feet by $\dfrac{12 \text{ inches}}{1 \text{ foot}}$

$$3.5 \text{ feet} \times \frac{12 \text{ inches}}{1 \text{ foot}} = \text{inches}$$

$$\frac{3.5 \times 12}{1} \text{ inches} = 42 \text{ inches}$$

In the above example we saw that, by multiplication and a subsequent division, 12 inches/1 foot converts 3.5 feet into 42 inches. Ratios, such as 12 inches/1 foot, are called conversion factors because they convert the units of a given quantity into the units and quantity of the unknown.

As you recall from Chapter One, the term "per" is the same as indicating a division process. For example:

Examples: 24 bottles per case means $\dfrac{24 \text{ bottles}}{1 \text{ case}}$ and is equivalent

to $\dfrac{1 \text{ case}}{24 \text{ bottles}}$ or $\dfrac{0.041 \text{ case}}{1 \text{ bottle}}$

6.02×10^{23} molecules in 22.4 liters means 6.02×10^{23}

molecules per 22.4 liters, or $\dfrac{6.02 \times 10^{23} \text{ molecules}}{22.4 \text{ liters}}$,

or $\dfrac{2.67 \times 10^{22} \text{ molecules}}{1 \text{ liter}}$, and is equivalent to

$\dfrac{1 \text{ liter}}{2.67 \times 10^{22} \text{ molecules}}$

20 miles per hour means $\dfrac{20 \text{ miles}}{1 \text{ hour}}$ or $\dfrac{1 \text{ mile}}{0.05 \text{ hour}}$, and

is equivalent to $\dfrac{0.05 \text{ hour}}{1 \text{ mile}}$

So we see, the "per" or "in" means the two terms (bottles and cases, miles and hours, etc.) are a ratio and that a change in the numerical value of one brings a corresponding numerical change in the other. Because the numerator and denominator are equivalent, they can be reversed and the ratio will still equal 1. The form of the ratio you choose will depend on what units you want to divide out.

PROBLEM. There are 24 bottles in a case of beer. If you have three cases, how many bottles of beer do you have? (Use a separate sheet of paper for your calculation steps.)

- - - - - - - - - - - - - -

Unknown: Number of bottles
Given: 24 bottles/1 case; 3 cases
Method: Dimensional analysis; convert 3 cases into bottles using the appropriate conversion factor.
Solution: Multiply 3 cases by a factor that will allow us to divide out the "case" units and leave bottles in the numerator equal to the unknown.

$3 \text{ cases} \times \dfrac{24 \text{ bottles}}{1 \text{ case}} = \text{bottles}$

$\dfrac{3 \times 24}{1} \text{ bottles} = 72 \text{ bottles}$

(5) PROBLEM. There are 6.02×10^{23} molecules of hydrogen in 22.4 liters of the gas. How many liters will hold 3.01×10^{23} molecules? (Don't worry if the terms "molecules" and "liters" are unfamiliar. You need not know what they mean to be able to solve the problem.)

- - - - - - - - - - - - - -

Unknown: Liters

Given: $\dfrac{6.02 \times 10^{23} \text{ molecules}}{22.4 \text{ liters}}$; 3.01×10^{23} molecules

Method: Dimensional analysis; convert 3.01×10^{23} molecules into liters

Solution: (1)   Write 3.01×10^{23} molecules.
(2)   Multiply by a conversion factor that will divide out molecules and give a product in terms of liters.

$$3.01 \times 10^{23} \text{ molecules} \times \frac{22.4 \text{ liters}}{6.02 \times 10^{23} \text{ molecules}}$$

$$\frac{(3.01 \times 10^{23})(22.4 \text{ liters})}{(6.02 \times 10^{23})} = \text{liters}$$

$$(0.5)(22.4 \text{ liters}) = 11.2 \text{ liters}$$

(6) PROBLEM. If one quart of a water sample contains 4×10^{10} bacteria, how many bacteria are contained in a gallon of the water? (Hint: *Two* conversions are necessary, because the given units differ.)

- - - - - - - - - - - - - -

Unknown: Number of bacteria

Given: $\dfrac{4 \times 10^{10} \text{ bacteria}}{1 \text{ quart}}$; one gallon

Method: Dimensional analysis — but a little more tricky, as we have to make two conversions, gallon to quarts and quarts to bacteria.

Solution: (1)   Write 1 gallon
(2)   Convert gallon to quarts with the appropriate conversion factor. (1 gallon = 4 quarts)

$$1 \text{ gallon} \times \frac{4 \text{ quarts}}{1 \text{ gallon}}$$

(3)   Now convert quarts to bacteria with the appropriate conversion factor.

$$1 \text{ gallon} \times \frac{4 \text{ quarts}}{1 \text{ gallon}} \times \frac{4 \times 10^{10} \text{ bacteria}}{1 \text{ quart}} = \text{bacteria}$$

$$\frac{1 \times 4 \times 4 \times 10^{10} \text{ bacteria}}{1 \times 1} = 16 \times 10^{10} \text{ bacteria, or}$$

$$1.6 \times 10^{11} \text{ bacteria}$$

(7)   Sometimes we find a problem that asks us to find a ratio of two or more units. In such a problem, the units in the denominator do not divide out as they did in previous problems.

PROBLEM. An automobile travels 60 miles in two hours. What is its speed in miles per hour?

– – – – – – – – – – – – –

Unknown: $\dfrac{\text{miles}}{\text{1 hour}}$

Given: 60 miles; two hours

Method: Dimensional analysis; the answer demands the units miles/1 hour

Solution: (1)   Write the miles traveled (since we want "miles" in the numerator).

            (2)   Multiply by $\dfrac{1}{\text{2 hours}}$ (since we want "hours" in the denominator).

$$60 \text{ miles} \times \frac{1}{2 \text{ hours}} = \frac{\text{miles}}{\text{hour}}$$

            (3) $\dfrac{60 \text{ miles}}{2 \text{ hours}} = \dfrac{30 \text{ miles}}{1 \text{ hour}}$

(8)   With this problem, do not be frustrated if you do not know the terms. It does not matter what the terms mean at this point. If you understand problem-solving procedures, you can do it.

PROBLEM. It takes 20 calories of heat to raise the temperature of 2 grams of water 10°C. What is the specific heat of water, if specific heat has the dimensions of $\dfrac{\text{calories}}{(1 \text{ gram})(1 \text{ degree})}$? (Note: Right now, we have $\dfrac{20 \text{ calories}}{(2 \text{ grams})(10 \text{ degrees})}$, which is not specific heat because 2 and 10 are in the denominator instead of 1 and 1.)

– – – – – – – – – – – – –

Unknown: $\dfrac{\text{calories}}{\text{1 gram} \cdot \text{1 degree}}$

Given: 20 calories; 2 grams; 10 degrees

Method: Dimensional analysis; the answer demands that calories be in the numerator and the product of grams and degrees be in the denominator.

Solution: (1)   Write 20 calories (because we want calories in the numerator).

                 20 calories

(2) Multiply by $\dfrac{1}{2 \text{ grams}}$ (because we want grams in the denominator).

(3) Multiply by $\dfrac{1}{10 \text{ degrees}}$ (because we want degrees in the denominator).

$$20 \text{ calories} \times \frac{1}{2 \text{ grams}} \times \frac{1}{10 \text{ deg.}} = \frac{1}{1 \text{ gram} \cdot 1 \text{ degree}}$$

(4) $\dfrac{20 \times 1 \times 1 \text{ calories}}{2 \times 10 \text{ grams} \cdot \text{degrees}} = \dfrac{1 \text{ calorie}}{1 \text{ gram} \cdot 1 \text{ degree}}$

METRIC CONVERSIONS

9 The following conversion factors and prefixes will be valuable when we solve problems using dimensional analysis throughout our study of the sciences — and increasingly in everyday life. We should be completely familiar with them.

Conversion Factors and Prefixes Within Metric System

PREFIX	ABBREVIATIONS	DEFINITION (times greater or lesser than the standard unit)
Mega	M	$1,000,000; \ 1 \times 10^6$
Kilo	k	$1,000; \ 1 \times 10^3$
Standard Unit	. . .	1.000
Deci	d	$1/10; 0.1; 1 \times 10^{-1}$
Centi	c	$1/100; 0.01; 1 \times 10^{-2}$
Milli	m	$1/1000; \ 0.001; 1 \times 10^{-3}$
Micro	μ	$1/1,000,000; 0.000001$ 1×10^{-6}

The *standard unit* is the basic unit of measure. It varies according to what is being measured. It was originally chosen somewhat arbitrarily but today is standardized throughout the world and held constant.

In the metric system, the standard unit of length is the meter, abbreviated m.

1 megameter (Mm) = 1,000,000 m
1 kilometer (km) = 1,000 m
1 decimeter (dm) = 0.1 m
1 centimeter (cm) = 0.01 m

1 millimeter (mm) = 0.001 m

1 micrometer (μm) = 0.000001 m

Another unit of length in the metric system is the Angstrom, abbreviated A and equal to 1×10^{-10} m.

The standard unit of volume in the metric systems is the liter, abbreviated l.

1 milliliter (ml) = 0.001 l

Also 1 milliliter = 1 cubic centimeter or 1 ml = 1 cm^3 or 1 ml = 1 cc

The standard unit of mass in the metric system is the gram, abbreviated g.

1 milligram (mg) = 0.001 g

These equivalents can then be used to form conversion factors for dimensional analysis problems in metric units.

PROBLEM. How many millimeters are there in 2.576 meters?

_ _ _ _ _ _ _ _ _ _ _ _ _ _

Unknown:	Millimeters (mm)
Given:	2.576 meters (m)
Method:	Dimensional analysis; 1000 mm = 1 m, or $\dfrac{1000 \text{ mm}}{1 \text{ m}}$

Solution: (1) $2.576 \text{ m} \times \dfrac{1000 \text{ mm}}{1 \text{ m}}$

(2) $\dfrac{2.576 \times 1000 \text{ mm}}{1} = 2{,}576 \text{ mm}$, or 2.576×10^3 mm

(10) PROBLEM. How many kilograms does 3.0 milligrams represent?

_ _ _ _ _ _ _ _ _ _ _ _ _ _

Unknown:	Kilograms (kg)
Given:	3.0 milligrams (mg)
Method:	Dimensional analysis; 1 kg = 1000 g or $\dfrac{1 \text{ kg}}{1000 \text{ g}}$;

1000 mg = 1 g or $\dfrac{1000 \text{ mg}}{1 \text{ g}}$ or $\dfrac{1 \text{ g}}{1000 \text{ mg}}$

(Remember you can reverse the numerator and denominator, which are equal, in forming conversion factors.)

Solution: (1) $3.0 \text{ mg} \times \dfrac{1 \text{ g}}{1000 \text{ mg}}$

(2) $3.0 \text{ mg} \cdot \dfrac{1 \text{ g}}{1000 \text{ mg}} \times \dfrac{1 \text{ kg}}{1000 \text{ g}} = \text{kg}$

(3) $\dfrac{3.0 \times 1 \times 1 \text{ kg}}{1000 \times 1000} = \dfrac{3.0 \text{ kg}}{1 \times 10^6} = 3.0 \times 10^{-6} \text{ kg}$

11 Here's a more complicated problem. Follow the solution carefully.

Example: A box measures 3 cm by 500 mm by 0.2 m. What is the volume of the box in liters?

Unknown: Volume in liters

Given: 3 cm by 500 mm by 0.2 m

Method: Dimensional analysis; the formula for the volume of a box: volume = length x width x height ($v = l \times w \times h$); convert units of length into cm, as cm^3 is easily related to liters; 1 m = 1000 mm or 1 m/10^3 mm; 1 m = 100 cm or 100 cm/1 m; 1000 cm^3 = 1 l, or (1 l/10^3 cm^3).

Solution: (1) $500 \text{ mm} \times \dfrac{1 \text{ m}}{10^3 \text{ mm}}$

(Note: For clarity, parentheses are used to separate length, width, and height dimensions.)

(2) $l = \left(500 \text{ mm} \times \dfrac{1 \text{ m}}{10^3 \text{ mm}} \times \dfrac{100 \text{ cm}}{1 \text{ m}} \right)$

(Note: The first conversion factor $\dfrac{1 \text{ m}}{10^3 \text{ mm}}$ changes 500 mm to meters and the second conversion factor $\dfrac{100 \text{ cm}}{1 \text{ m}}$ changes it to centimeters which are needed for the problem.)

(3) $l \times w = \left(500 \text{ mm} \times \dfrac{1 \text{ m}}{10^3 \text{ mm}} \times \dfrac{100 \text{ cm}}{1 \text{ m}} \right)$

$\times \left(0.2 \text{ m} \times \dfrac{100 \text{ cm}}{1 \text{ m}} \right)$

(Note: $\dfrac{100 \text{ cm}}{1 \text{ m}}$ was used to convert 0.2 m to cm.)

(4) $l \times w \times h = \left(500 \text{ mm} \times \dfrac{1 \text{ m}}{10^3 \text{ mm}} \times \dfrac{100 \text{ cm}}{1 \text{ m}} \right)$

$\times \left(0.2 \text{ m} \times \dfrac{100 \text{ cm}}{1 \text{ m}} \right) \times (3 \text{ cm})$

(Note: Height does not have to be converted as it is already in centimeters.)

(5) At this point the only remaining unit is cm x cm x cm = cm^3, so we use the conversion factor $\dfrac{1 \text{ l}}{10^3 \text{ cm}^3}$ to convert cm^3 to liters.

$\left(500 \text{ mm} \times \dfrac{1 \text{ m}}{10^3 \text{ mm}} \times \dfrac{100 \text{ cm}}{1 \text{ m}} \right)$

$\times \left(0.2 \text{ m} \times \dfrac{100 \text{ cm}}{1 \text{ m}} \right) \times (3 \text{ cm}) \times \dfrac{1 \text{ l}}{10^3 \text{ cm}^3}$

(6)   We can then rewrite the numerical part of the equation and calculate our answer in liters.

$$\frac{500 \times 1 \times 100 \times 0.2 \times 100 \times 3}{10^3 \times 1 \times 1 \times 10^3} \text{ liter} = \text{liters}$$

(7) $\dfrac{300 \times 10^4}{1 \times 10^6}$ liter = 300×10^{-2} liters, or 3.0 liters

PROBLEM. A balloon is 10 cm in diameter. What is the volume of the balloon in cubic meters? (Note: The equation for the volume of a sphere is $V = \dfrac{4\pi r^3}{3}$ where r (radius) = $\dfrac{\text{diameter}}{2}$.)

_ _ _ _ _ _ _ _ _ _ _ _ _ _ _

Unknown: Volume in cubic meters
Given: Diameter = 10 centimeters
Method: Dimensional analysis

(1)   Formula: $V = \dfrac{4\pi r^3}{3}$

(2)   Convert diameter to r.
$$r = \frac{10 \text{ cm}}{2} = 5 \text{ cm}$$

(3)   Convert cm to m.
100 cm = 1 m
$$1 = \frac{1 \text{ m}}{100 \text{ cm}}$$

Solution: (1) $r = (5 \text{ cm}) \left(\dfrac{1 \text{ m}}{100 \text{ cm}} \right) = .05 \text{ m}$

(2) $V = \dfrac{4\pi(.05)^3}{3} = \dfrac{4\pi(5 \times 10^{-2}\text{m})^3}{3} =$

$\dfrac{4\pi(125 \times 10^{-6}\text{m}^3)}{3} = \dfrac{500\pi \times 10^{-6}}{3} \text{m}^3$

CONVERSION BETWEEN THE METRIC SYSTEM AND THE ENGLISH SYSTEM

⑫   To convert from the metric system to the English system and vice versa, you should know at least one conversion factor for length, volume, and mass. We assume you know the English system (how many quarts in a gallon? yards in a mile?, etc.). If you feel you don't, review now. A table of the English system is given in the Appendix. With these conversion factors, you use the familiar method of dimensional analysis.

Length: 2.54 centimeters = 1 inch   or   2.54 cm/in
          1 meter = 39.37 inches   or   1 m/39.37 in

Volume: 1 liter = 1.06 qt   or   1 l/1.06 qt

            1 qt = 0.9463 l   or   1 qt/0.9463 l

Mass: 1 kilogram = 2.20 lbs   or   1 kg/2.20 lbs

            1 pound = 454 grams   or   1 lb/454 g

PROBLEM.   Convert 14 lbs to grams. _____

- - - - - - - - - - - - - - -

Unknown: grams (g)

Given: 14 pounds (lb)

Method: Dimensional analysis; 1 lb = 454 g, or 454 g/1 lb. (Remember, we can flip numerator and denominator to get the units correct.)

Solution: (1) $14 \text{ lbs} \times \dfrac{454 \text{ g}}{1 \text{ lb}} = \text{g}$

            (2) $\dfrac{14 \times 454 \text{ g}}{1} = 6,356 \text{ g}$

(13)   PROBLEM.   How many feet are there in two meters?

_____

- - - - - - - - - - - - - - -

Unknown: feet (ft)

Given: 2 meters (m)

Method: Dimensional analysis; 1 m = 39.37 in, or 39.37 in/1 m; 12 in = 1 ft, or 12 in/1 ft, or 1 ft/12 in.

Solution: (1) $2 \text{ m} \times \dfrac{39.37 \text{ in}}{1 \text{ m}}$ (Convert m to in.)

            (2) $2 \text{ m} \times \dfrac{39.37 \text{ in}}{1 \text{ m}} \times \dfrac{1 \text{ ft}}{12 \text{ in}} = \text{ft}$ (Convert in to ft after converting m to in.)

            (3) $\dfrac{2 \times 39.37 \times 1}{1 \times 12} \text{ ft} = 6.56 \text{ ft}$

(14)   PROBLEM.   How many gallons are there in 420 cm^3?

_____

- - - - - - - - - - - - - - -

Unknown: gallons (gal)

Given: 420 cm^3

Method: Dimensional analysis; 1 gal/4 qt; 1.06 qt/1 l; 10^{3}cm^3/1 l

Solution: (1) $420 \text{ cm}^3 \times \dfrac{1 \text{ l}}{10^3 \text{ cm}^3}$

            (2) $420 \text{ cm}^3 \times \dfrac{1 \text{ l}}{10^3 \text{ cm}^3} \times \dfrac{1.06 \text{ qt}}{1 \text{ l}}$

(3) $420 \text{ cm}^3 \times \dfrac{1 \text{ l}}{10^3 \text{ cm}^3} \times \dfrac{1.06 \text{ qt}}{1 \text{ l}} \times \dfrac{1 \text{ gal}}{4 \text{ qt}} = \text{gal}$

(4) $\dfrac{420 \times 1 \times 1.06 \times 1}{10^3 \times 1 \times 4} \text{ gal} = 0.11 \text{ gal}$

---

15 PROBLEM. Beer in Germany is usually sold by the liter stein, while in America it is usually sold by the 12 oz bottle. How many 12 oz bottles are in a 1 liter stein?

_____

– – – – – – – – – – – – –

Unknown: bottles
Given: 1 liter; 12 oz/1 bottle (or 1 bottle/12 oz)
Method: Dimensional analysis; 1 l/1.06 qt; 32 oz/1 qt
Solution: (1) $1 \text{ l} \times \dfrac{1.06 \text{ qt}}{1 \text{ l}}$ (Convert liters to quarts.)

         (2) $1 \text{ l} \times \dfrac{1.06 \text{ qt}}{1 \text{ l}} \times \dfrac{32 \text{ oz}}{1 \text{ qt}}$ (Convert quarts to ounces.)

         (3) $1 \text{ l} \times \dfrac{1.06 \text{ qt}}{1 \text{ l}} \times \dfrac{32 \text{ oz}}{1 \text{ qt}} \times \dfrac{1 \text{ bottle}}{12 \text{ oz}} = \text{bottles}$
         (Convert ounces to bottles.)

         (4) $\dfrac{1 \times 1.06 \times 32 \times 1}{1 \times 1 \times 12}$ bottles = 2.82 bottles

---

16 PROBLEM. One atom of carbon weighs 1.99×10^{-23} grams. How many pounds does it weigh?

_____

– – – – – – – – – – – – –

Unknown: pounds (lbs)
Given: 1.99×10^{-23} g; 1 atom of carbon
Method: Dimensional analysis; 1 lb/454 g
Solution: (1) $\dfrac{1.99 \times 10^{-23} \text{ g}}{1 \text{ atom of carbon}} \times \dfrac{1 \text{ lb}}{454 \text{ g}} = \text{lb}$

         (2) $\dfrac{1.99 \times 10^{-23} \times 1}{1 \text{ atom of carbon} \times 454} \text{ lb} = \dfrac{4.38 \times 10^{-26} \text{ lb}}{1 \text{ atom of carbon}}$

MORE ON CONCEPTS AND DIMENSIONAL ANALYSIS

17 Often dimensional (or units) analysis is not enough to solve problems. Many problems in science demand judicious application of concepts and definitions. The next few frames will define several concepts and will demonstrate how to solve problems by uniting concepts with dimensional analysis. The first three concepts — percent, density, and temperature conversion — have wide application and should be useful to all readers. The last two concepts — heat change and heat transfer — will be most useful for students studying chemistry or chemistry-related courses.

Percentage

Percentage is used in most sciences and in everyday life as well. It is one concept with which everyone should be familiar. For instance, often we are given a percent and asked to find the portion it represents.

Definition: $\text{Percent} = \dfrac{\text{Part}}{\text{Whole}} \times 100\% \text{ whole}$

Example: If there are 300 students in your class and 270 of them are female, what percent of the class is female? What percent is male?

Unknown: % female; % male
Given: Class of 300 students; number of females = 270
Method: Dimensional analysis and formula. The sum of the percent female plus the percent male should equal 100%.
Solution: (1) Use formula and units.

$$\frac{\text{no. of females}}{\text{no. of students}} \times 100\% \text{ students} = \% \text{ females}$$

(2) Substitute values for "part" and "whole."

$$\frac{270 \text{ females}}{300 \text{ students}} \times 100\% \text{ students} = \% \text{ females}$$

(3) $\dfrac{270 \times 100}{300} \%$ females = 90% females

(4) Percent males will be equal to the difference between 100% and percent females.
% males = 100% − 90% = 10% males

PROBLEM. A tall pea plant was crossed with a short pea plant. Twenty offspring were produced. Eight were short. What percent of the offspring were short?

— — — — — — — — — — — —

Unknown: % short
Given: 20 offspring; 8 short
Method: Dimensional analysis and formula
Solution: (1) Use formula and units.

$$\frac{\text{no. of shorts}}{\text{no. of offspring}} \times 100\% \text{ offspring} = \% \text{ shorts}$$

(2) Substitute values.

$$\frac{8 \text{ shorts}}{\underset{1}{\cancel{20}}} \times \cancel{100}^{5}\% = 40\% \text{ shorts}$$

(18) To find a part, given the percent and the whole, we use the same formula.

Example: If a particular alloy is 8.5% zinc, how many grams of zinc are contained in 50 grams of the alloy?

Unknown: zinc (grams)

Given: 8.5% zinc, 50 g of alloy

Method: Dimensional analysis; here we are finding how much "part" we have in a "whole" of 50 g.

Solution: (1) Since the unit "zinc" is going to appear in our answer, we write % zinc.

8.5% zinc

(2) Multiply by 1/(100% alloy) to divide out "%" term and introduce the "alloy" unit.

$$8.5\cancel{\%} \text{ zinc} \times \frac{1}{100\cancel{\%} \text{ alloy}}$$

(3) Multiply by 50 g alloy, to divide out "alloy" term and introduce "gram" unit.

$$8.5\cancel{\%} \text{ zinc} \times \frac{1}{100\cancel{\%} \text{ alloy}} \times 50 \text{ g } \cancel{\text{alloy}} = \text{g zinc}$$

(4) $$\frac{8.5 \times 1 \times 50 \text{ g zinc}}{100} = 4.25 \text{ g zinc}$$

(Note: You could have rearranged the formula so that the "part" was isolated instead of using dimensional analysis.)

PROBLEM. John scored higher than 98% of students his age on an achievement test. If 300,000 students took the test, how many students scored lower than John?

- - - - - - - - - - - - - -

Unknown: # of lower students

Given: 98% lower students; 300,000 total

Method: Formula

Solution: (1) $$\% \text{ lower} = \frac{\text{\# lower}}{\cancel{\text{total}}} \times 100\% \cancel{\text{ total}}$$

(2) Substitute values.

$$98\% \text{ lower} = \frac{\text{\# lower}}{300,000} \times 100\%$$

(3) Multiply by $\frac{300,000}{100\%}$

$$\frac{98\% \text{ lower} \times 300,000}{100\%} = \text{\# lower}$$

$$294,000 \text{ lower} = \text{\# lower}$$

⑲   See how you do with this one. It gives you the part and the percent and you must find the whole.

PROBLEM. Air is 20% oxygen by weight. How many grams of air will contain 15 grams of oxygen?

— — — — — — — — — — — — — —

Unknown: grams of air
Given: 20% oxygen; 15 g oxygen
Method: Here we are given percent and part; we must calculate the whole.
Solution: (1) To calculate the whole (g air), we write grams of oxygen so that grams will appear in the numerator.

15 g oxygen

(2) Multiply by $\dfrac{1}{20\% \text{ oxygen}}$ so that the oxygen term will divide out.

$15 \text{ g oxygen} \times \dfrac{1}{20\% \text{ oxygen}}$

(3) Multiply by 100% air so that percent divides out.

$15 \text{ g oxygen} \times \dfrac{1}{20\% \text{ oxygen}} \times 100\% \text{ air}$

(4) $\dfrac{15 \times 1 \times \overset{5}{100}}{\underset{}{20}}$ grams air = 75 grams air

Density

⑳   Density is important in the sciences but it is also being heard more in everyday speech (e.g., population density).

Definition: Density is the ratio of mass to volume;

$$\text{Density} = \dfrac{\text{mass}}{\text{volume}}$$

Example: Five milliliters of water weighs 5 grams. What is the density of water in grams per milliliter?

Unknown: density (g/ml)
Given: 5 ml; 5 g
Method: Dimensional analysis
Solution: (1) Write 5 g (We want grams in the numerator; the answer demands it.)

5 g

(2) Multiply by $\dfrac{1}{5 \text{ ml}}$ (We want milliliter in the denominator; the answer demands it.)

$$5 \text{ g} \times \frac{1}{5 \text{ ml}} = \frac{\text{g}}{\text{ml}}$$

(3) $\dfrac{5 \text{ g}}{5 \text{ ml}} = 1 \text{ g/ml}$

(Note: The formula would have given us the same thing since 5 g = mass and 5 ml = volume.)

PROBLEM. If copper has a density of 8.9 g/cm^3 and a block of copper weighs 17.8 g, what is the volume occupied by the copper?

— — — — — — — — — — — —

Unknown: cm^3 of copper (volume)

Given: $\dfrac{8.9 \text{ g}}{\text{cm}^3}$ (density); 17.8 g of copper (mass)

Method: Formula

Solution: (1) Rearrange formula.

$$(\text{volume})(\text{density}) = \frac{\text{mass}}{\cancel{\text{volume}}} \cdot \cancel{\text{volume}}$$

$$\frac{(\text{volume})(\cancel{\text{density}})}{\cancel{\text{density}}} = \frac{\text{mass}}{\text{density}}$$

$$\text{volume} = \frac{\text{mass}}{\text{density}}$$

(2) Substitute values.

$$\text{volume} = \frac{17.8 \text{ g}}{8.9 \text{ g/cm}^3}$$

$$\text{volume} = \frac{17.8}{8.9} \frac{\cancel{\text{(g)}} (\text{cm}^3)}{\cancel{\text{g}}}$$

$$\text{volume} = 2 \text{ cm}^3$$

(21) PROBLEM. Fifteen people occupy an area that is 200 feet square. What is the population density? (Note: The density formula is changed somewhat in that mass becomes # of people and volume becomes area.)

— — — — — — — — — — — —

Unknown: Density (people/square feet)
Given: 15 people; square with 200-foot sides.
Method: Dimensional analysis
Solution: (1) "15 people" forms the numerator.
          (2) "area" forms the denominator.
          (3) $\dfrac{15 \text{ people}}{\text{area}} = \dfrac{15 \text{ people}}{200 \times 200 \text{ sq. ft.}}$

$= 3.75 \times 10^{-4}$ people/sq. ft., or 1 person/2,667 sq. ft.

(22) PROBLEM. If a 5% (by weight) sugar solution has a density of 1.5 g/ml, how many grams of sugar are found in 100 ml of solution? (Hint: To solve this problem you will need to use percents and density.)

- - - - - - - - - - - - - - -

Unknown: grams sugar

Given: density = 1.5 g/ml; 100 ml; 5% solution

Method: Find total grams and then convert percent composition by dimensional analysis.

Solution: (1) Write density with grams in the denominator — because we want grams in the answer.

$$\frac{1.5 \text{ g}}{\text{ml}}$$

(2) Multiply by 100 ml solution to get total weight of solution.

$$\frac{1.5 \text{ g}}{\text{ml}} \times 100 \text{ ml solution} = 150 \text{ grams of solution}$$

(3) Now, multiply by $\dfrac{1}{100\% \text{ solution}}$ to eliminate the solution term and introduce %.

$$150 \text{ grams of solution} \times \frac{1}{100\% \text{ solution}}$$

(4) Multiply by 5% sugar.

$$150 \text{ grams of solution} \times \frac{1}{100\% \text{ solution}} \times 5\% \text{ sugar}$$

(5) $\dfrac{150 \times 1 \times 5 \text{ g sugar}}{100} = 7.5 \text{ g sugar}$

(23) PROBLEM. Given the following data, and assuming the metal rod is completely immersed in water, calculate the density (in g/ml) of the metal rod. (Hint: The mass and volume of the rod are given indirectly. Find each before you try to find the density.)

Volume of water in graduated cylinder = 15.2 ml
Weight of graduated cylinder + water = 97.5 g
Volume of water + metal rod = 19.3 ml
Weight of graduated cylinder + water + metal rod = 104.0 g

- - - - - - - - - - - - - - -

Unknown: density of metal rod (g/ml)

Given: mass and volume of the metal rod (indirectly)

Method: Evaluate the mass and volume of the rod in separate steps, then calculate density through dimensional analysis.

Solution: (1) Evaluate weight of rod [(wt. of grad. cyl. + H_2O + rod) − (wt. of grad. cyl. + H_2O)].

104.0 g − 97.5 g = 6.5 g

(2) Evaluate volume of rod.

19.3 ml − 15.2 ml = 4.1 ml

(3) Evaluate density of rod.

$$6.5 \text{ g} \times \frac{1}{4.1 \text{ ml}} = \text{g/ml}$$

$$\frac{6.5 \text{ g}}{4.1 \text{ ml}} = 1.6 \text{ g/ml}$$

Temperature Conversion

24 Here are the temperature conversion formulas that you should memorize for both scientific and everyday problems.

$$°C = \frac{(°F - 32°)}{1.8} \text{ where }°C \text{ means degrees Centigrade}$$

$$°F = 1.8°C + 32° \text{ where }°F \text{ means degrees Fahrenheit}$$

$$°K = °C + 273° \text{ where }°K \text{ means degrees Kelvin or absolute}$$

Example: Normal room temperature is 75°F. What temperature does this correspond to on the Centigrade scale?

Unknown: °C
Given: 75°F
Method: $°C = \dfrac{(°F - 32°)}{1.8}$

Solution: (1) Substitute 75° for °F in the equation.

$$°C = \frac{(75° - 32°)}{1.8}$$

(2) $°C = \dfrac{43°}{1.8}$

$$°C = 23.9°$$

PROBLEM. If 68°F is the room temperature setting used by federal buildings to conserve energy, what is the °C setting?

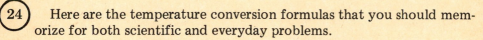

Unknown: °C
Given: 68°F
Method: $°C = \dfrac{(°F - 32°)}{1.8}$
Solution: (1) Substitute 68° for °F in the equation.

$$°C = \frac{68° - 32°}{1.8}$$

$$(2) \quad °C = \frac{36°}{1.8}$$

$$°C = 20°$$

25 PROBLEM. The temperature at which a substance melts is 18°F. At what temperature does the substance melt on the Kelvin or Absolute temperature scale?

_ _ _ _ _ _ _ _ _ _ _ _ _ _ _

Unknown: °K

Given: 18°F

Method: First convert °F to °C and then °C to °K.

$$°C = \frac{(°F - 32°)}{1.8} ; \quad °K = °C + 273°$$

Solution: (1) $°C = \dfrac{(18° - 32°)}{1.8}$

(2) °K = °C + 273° (Substitute the whole expression for degrees Centigrade into the equation defining °K.)

$$°K = \frac{(18° - 32°)}{1.8} + 273°$$

(3) $°K = \dfrac{-14°}{1.8} + 273°$

$$°K = -7.8° + 273°$$

$$°K = 265.2°$$

Heat Change

26 Heat change and heat transfer are primarily topics for chemistry students. If these topics have no application for you at this time, skip ahead to frame 31 on estimating.

Definitions: 1 calorie (cal.) is the amount of heat energy required to raise the temperature of 1 g of water by 1°C.

When a substance is warmed, the change in heat energy is said to be positive. When a substance is cooled, the change in heat energy is said to be negative or less than zero.

Specific heat of a substance =

$$\frac{\text{change in heat energy}}{\text{weight} \times \text{change in temperature}}$$

Temperature change, often symbolized ΔT, is defined as final temperature minus initial temperature.

Specific heat has the dimensions: $\dfrac{\text{calories}}{\text{g} \cdot \text{deg.}}$

Specific heat of water is $\dfrac{1 \text{ cal}}{1 \text{ g} \cdot 1 \text{ deg.}}$

Example: If 70 calories of heat are required to raise the temperature of 32 grams of a metal from 25.0°C to 45.6°C, what is the specific heat of the metal?

Unknown: Specific heat of a metal: $\dfrac{1 \text{ cal.}}{1 \text{ g} \cdot 1 \text{ deg.}}$

Given: Calories = 70 cal.; weight of the metal = 32 g; and temperature change from 25.0°C to 45.6°C.

Method: Find change in temperature and then apply dimensional analysis.

Solution:
(1) $\Delta T = 45.6°C - 25.0°C$
$\Delta T = 20.6°C$

(2) Dimensional analysis — write calories.

70 cal.

(3) Multiply by the reciprocal of 32 g (we want grams in the denominator).

$70 \text{ cal.} \times \dfrac{1}{32 \text{ g}}$

(4) Multiply by the reciprocal of ΔT (we want degrees in the denominator).

$70 \text{ cal.} \times \dfrac{1}{32 \text{ g}} \times \dfrac{1}{20.6°C} = \dfrac{\text{cal.}}{\text{g} \cdot \text{deg.}}$

$\dfrac{70 \text{ cal.}}{32 \times 20.6 \text{ g} \cdot \text{deg.}} = 0.106 \dfrac{\text{cal.}}{\text{g} \cdot \text{deg.}}$

PROBLEM. How many calories of heat are required to raise the temperature of 20 g of water by 20°C?

— — — — — — — — — — — — — —

Unknown: Number of calories
Given: 20 g water; $\Delta T = 20°C$
Method: Dimensional analysis; specific heat of water = 1 cal/g · deg
Solution:
(1) Write specific heat of water (this puts cal. in the numerator as is demanded by the answer).

$\dfrac{1 \text{ cal.}}{\text{g} \cdot \text{deg.}}$

(2) Multiply by 20 g (this allows us to divide out the g term).

$\dfrac{1 \text{ cal.}}{\cancel{\text{g}} \cdot \text{deg.}} \times 20 \cancel{\text{g}}$

(3) Multiply by 20 (to divide out deg. term).

$$\frac{1 \text{ cal.}}{\cancel{g} \cdot \cancel{deg.}} \times 20 \cancel{g} \times 20 \cancel{deg.} = \text{cal.}$$

1 × 20 × 20 cal. = 400 cal.

27 PROBLEM. How many calories of heat are lost when 14 g of iron (specific heat of 0.107 cal/g · deg.) is cooled from 47°C to 25°C?

– – – – – – – – – – – –

Unknown: Number of calories
Given: 14 g iron; specific heat = 0.107 cal./g · deg.; temperature change from 47°C to 25°C.
Method: Dimensional analysis
Solution: (1) Write specific heat (we want calories in the numerator as defined by the unknown).

$$\frac{0.107 \text{ cal.}}{g \cdot deg.}$$

(2) Multiply by number of grams (to divide out grams).

$$\frac{0.107 \text{ cal.}}{\cancel{g} \cdot deg.} \times 14 \cancel{g}$$

(3) Multiply by ΔT (final temperature minus initial temperature) to eliminate degrees.

$$\frac{0.107 \text{ cal.}}{\cancel{g} \cdot deg.} \times 14 \cancel{g} \times (25°C - 47°C)$$

$$\frac{0.107 \text{ cal.}}{\cancel{g} \cdot \cancel{deg.}} \times 14 \cancel{g} \times (-22 \cancel{deg.}) = \text{calories}$$

0.107 × 14 × (−22) cal. = −32.96 cal.

28 PROBLEM. If 10 g of water with an initial temperature of 30°C loses 15 calories of heat energy, what is the final temperature?

– – – – – – – – – – – –

Unknown: Final temperature
Given: 10 g water; initial temperature = 30°C; calorie loss = 15 cal.
Method: This problem can be solved two ways: (a) by algebra or (b) by dimensional analysis. Specific heat of water = 1 cal./1 g · deg.
Solutions:
(a) (1) Specific heat = $\dfrac{1 \text{ cal.}}{1 \text{ g} \cdot 1 \text{ deg.}}$

(2) Let Δcal. = change in calories; let M = weight of water; and let $(T_f - T_i)$ = change in temperature.

$$\frac{1 \text{ cal.}}{1 \text{ g} \cdot 1 \text{ deg.}} = \frac{\Delta \text{cal.}}{M(T_f - T_i)}$$

(3) Solve for T_f

$$T_f - T_i = \frac{\Delta \text{cal.} \times 1 \text{ g} \times 1 \text{ deg.}}{M \times 1 \text{ cal.}}$$ (Multiply both sides

by $\frac{(T_f - T_i)(1 \text{ g})(1 \text{ deg.})}{1 \text{ cal.}}$.)

$$T_f = \frac{\Delta \text{cal.} \times 1 \text{ g} \times 1 \text{ deg.}}{M \times 1 \text{ cal.}} + T_i$$ (Add T_i to both sides.)

(4) Substitute values for symbols and calculate.

Δcal. = −15 cal.

M = 10 g

T_i = 30°C

$$T_f = \frac{-15 \text{ cal.} \times 1 \text{ g} \times 1 \text{ deg.}}{10 \text{ g} \times 1 \text{ cal.}} + 30°C$$

$$T_f = \frac{-15}{10} \text{ deg.} + 30°C$$

$T_f = -1.5°C + 30°C$

$T_f = 28.5°C$

(b) (1) Solve for ΔT by dimensional analysis, that is, we want degrees in the numerator of the answer so we write the reciprocal of the specific heat of water.

$\frac{1 \text{ deg.} \cdot 1 \text{ g}}{1 \text{ cal.}}$

(2) Multiply by change in calories

$\frac{1 \text{ deg.} \cdot 1 \text{ g}}{1 \text{ cal.}} \times (-15 \text{ cal.})$

(3) Multiply by 1/10 g, to eliminate gram term.

$\frac{1 \text{ deg.} \cdot 1 \text{ g}}{1 \text{ cal.}} \times (-15) \text{ cal.} \times \frac{1}{10 \text{ g}}$ = deg. change

$\frac{-15}{10}$ = deg. change

−1.5 = deg. change

(4) Now solve ΔT, for T_f (we cannot get away from algebra).

$\Delta T = T_f - T_i$

$\Delta T + T_i = T_f$

−1.5°C + 30°C = T_f

28.5°C = T_f

It is important to realize that problems can be solved by a variety of approaches, and the strategies and tactics for problem solving cannot be taught, but learned. Learning can only come from repeated applications of definitions and concepts to problem solving.

Heat Transfer

(29) Heat change leads to heat transfer. Since energy cannot be created or destroyed, the heat lost by one body will equal the heat gained by another body.

The heat change for any substance is also defined as the product of its specific heat, its mass, and the change in temperature that the substance undergoes.

$$\text{heat change} = \text{specific heat} \times \text{mass} \times \Delta T$$

In units this is expressed as:

$$\text{cal.} = \frac{\text{cal.}}{\text{g} \cdot \text{deg.}} \times \text{g} \times \text{deg.}$$

When substances at two different temperatures are brought into contact, heat energy will flow from the warmer to the cooler until they reach the same temperature.

$$- [\text{heat lost}] = \text{heat gained}$$

or:

$$- [(\text{Sp. heat})_a \ (\text{mass})_a \ (\Delta T_a)] = (\text{Sp. heat})_b \ (\text{mass})_b \ (\Delta T)_b$$

Example: Two grams of a substance at 50°C is immersed in 50 ml of water which had an initial temperature of 25°C. If the equilibrium temperature of the water was 35°C, what is the specific heat of the substance?

Unknown: Specific heat of substance (cal./g · deg.)

Given: 2 g of substance; 50 ml of H_2O; initial temperature of substance = 50°C; initial temperature of water = 25°C; final temperature of water and substance = 35°C.

Method: Find the change in heat energy for water and then apply dimensional analysis to find specific heat. 50 ml of water is equivalent to 50 g of water — because water has a density of 1 g/cm³.

Solution: (1) Write specific heat of water (we want heat change for water).

$$\frac{1 \text{ cal.}}{\text{g} \cdot \text{deg.}}$$

(2) Multiply by weight of water (to eliminate gram term).

$$\frac{1 \text{ cal.}}{\text{g} \cdot \text{deg.}} \times 50 \text{ g}$$

(3) Multiply by ΔT for water (to eliminate degree term).

$$\Delta T = 35°C - 25°C = 10°C$$

$$\frac{1 \text{ cal.}}{g \cdot deg.} \times 50 \, g \times 10°C = \text{cal. gained by water}$$

$$1 \times 50 \times 10 = 500 \text{ cal.}$$

(4) Heat gained by water = heat lost by substance. Calories lost = −500 cal., write this term, as we want calorie in the numerator of the answer.

−500 cal.

(5) Multiply by 1/2 g, to get grams in the denominator.

$$-500 \text{ cal.} \times \frac{1}{2 \, g}$$

(6) Multiply by $1/\Delta T$ of substance, to get degrees in the denominator.

$\Delta T = 35°C - 50°C = -15°C$ (T_f is lower than T_i so Δ is negative.)

$$-500 \text{ cal.} \times \frac{1}{2 \, g} \times \frac{1}{-15°C} = \text{specific heat of}$$

substance

$$\frac{-500 \text{ cal.}}{-30 \, g \cdot deg.} = 16.7 \, \frac{\text{cal.}}{g \cdot deg.}$$

PROBLEM. A piece of hot iron weighing 5 grams is touched to a piece of cold iron weighing 10 grams. The hot iron has an initial temperature of 50°C and a final temperature of 25°C. What was the cold iron's initial temperature?

— — — — — — — — — —

Unknown: T_i for cold iron

Given: M_h = 5 grams; M_c = 10 grams; specific heat$_h$ = specific heat$_c$, since both are iron. $T_{i\text{-}h}$ = 50°C; $T_{f\text{-}h}$ = 25°C; $T_{f\text{-}c}$ = 25°C

Method: − [Heat change for hot iron] = heat change for cold iron (because hot iron's change is negative but cold iron's change is positive). Heat change = (specific heat)(mass)(change in temperature).

Solution: (1) Specific heat cancels out on both sides since they are equal.

(2) Substitute values

− [(5 g)(25° − 50°)] = (10 g)(25° − T_i)

(3) $-\left[\dfrac{5 \, g(25° - 50°)}{10 \, g} \right] = 25° - T_i$

+ 12.5° = 25° − T_i

T_i = 12.5°C

(30)   Here is an example problem which requires analysis in general terms of the physical situation — another technique of problem solving.

Example: If 30 g of water at 25°C is mixed with 40 g of water at 45°C, what will be the equilibrium temperature of the resultant 60 g of water?

Unknown: Final temperature (deg.)

Given: 30 g of water at 25°C; 40 g of water at 45°C

Method: Describe in general terms the physical event. Our reasoning should proceed like this:

(1)   The final temperature will be somewhere between 25°C and 45°C.

(2)   The change in temperature, ΔT, for the water at 25°C will be positive, that is, the temperature will increase.

(3)   We can write the ΔT for the 25°C water as $T_f - 25°C$.

(4)   The 45°C water will have a lower temperature.

(5)   Write ΔT for 45°C water as $45°C - T_f$.

A diagram of the situation might help:

45°C
$\left.\vphantom{\begin{array}{c}a\\b\end{array}}\right\}$ ΔT for 45°C water $= 45°C - T_f$
T_f
$\left.\vphantom{\begin{array}{c}a\\b\end{array}}\right\}$ ΔT for 25°C water $= T_f - 25°C$
25°C

Solution: Now we can set up an algebraic equation for heat lost = heat gained.

$$\left(\frac{1\ cal.}{g \cdot deg.}\right)(40\ g)(45°C - T_f) = \left(\frac{1\ cal.}{g \cdot deg.}\right)(30\ g)$$
$$(T_f - 25°C)$$

$(40 \times 45\ cal.) - \left(\dfrac{40\ cal.}{deg.}\right)T_f = \left(\dfrac{30\ cal.}{deg.}\right)T_f - (30 \times$ 25 cal.) (Note: The result of the multiplication $\left(\dfrac{1\ cal.}{\cancel{g} \cdot \cancel{deg.}}\right)(40\cancel{g})(45\cancel{°})$ is 40 × 45 cal., because deg. cancels with ° and g with g. The same thing happens with the other units. Remember that you must distribute $\left(\dfrac{1\ cal.}{g \cdot deg.}\right)(40\ g)$ to T_f also.)

$$1800\ cal. - \left(\frac{40\ cal.}{deg.}\right)T_f = \left(\frac{30\ cal.}{deg.}\right)T_f - 750\ cal.$$

$$1800\ cal. + 750\ cal. = \left(\frac{30\ cal.}{deg.}\right)T_f + \left(\frac{40\ cal.}{deg.}\right)T_f$$

$$2550\ cal. = \left(\frac{70\ cal.}{deg.}\right)T_f$$

$$\left(\frac{deg.}{70\ cal.}\right)(2550\ cal.) = T_f$$

$$36.4°C = T_f$$

PROBLEM. If 50 grams of iron (specific heat = $\frac{.107 \text{ cal.}}{g \cdot \text{deg.}}$) at 1000°C is suddenly immersed in 200 grams of water at 25°C, can you put your hand into the water without getting a bad burn? (Bad burns are caused by temperatures over 60°C.)

Unknown:	Indirect — final temperature of the water
Given:	50 g of iron; specific heat = .107 cal/g · deg.; initial temperature of iron = 1000°C; 200 g of water at 25°C.
Concepts:	− [Heat lost] = heat gained; heat = specific heat × mass × ΔT.
Solution:	$\left(\frac{.107 \text{ cal.}}{g \cdot \text{deg.}}\right)$ (50 g)(1000°C − T_f) = $\left(\frac{1 \text{ cal.}}{g \cdot \text{deg.}}\right)$ (200 g)

$(T_f - 25°C)$ (Note: The negative caused T_f − 1000°C to be switched.)

$$5{,}350 \text{ cal.} - \left(5.35 \frac{\text{cal.}}{\text{deg.}}\right) T_f = \left(200 \frac{\text{cal.}}{\text{deg.}}\right) T_f - 5000 \text{ cal.}$$

$$5{,}350 \text{ cal.} + 5{,}000 \text{ cal.} = \left(200 \frac{\text{cal.}}{\text{deg.}}\right) T_f + \left(5.35 \frac{\text{cal.}}{\text{deg.}}\right) T_f$$

$$10{,}350 \text{ cal.} = \left(205.35 \frac{\text{cal.}}{\text{deg.}}\right) T_f$$

$$\left(\frac{\text{deg.}}{205.35 \text{ cal.}}\right) (10{,}350 \text{ cal.}) = T_f$$

$$50.4°C = T_f$$

You can put your hand in, if you want to.

ESTIMATION

(31) Often in test problems, we are required to do the set-up only. That is, we need not actually do the arithmetic. A *set-up* is considered to be that point in our problem solving at which we have all of our known quantities lumped together on one side of the equation and our unknown quantity on the other. At this point, we would only need to carry out the arithmetic operations on numbers and units to get a numerical answer.

An important skill in working numerical problems to a set-up is the ability to estimate the approximate order of magnitude of the answer. We want to be able to tell quickly whether or not we are in the right ball park.

Estimation helps us check this quickly. Estimation is a two-step process: first we check units, then we check numbers. To estimate units, we first check the units of measure in our set-up; if these units do not correspond to the units demanded by the answer, then the set-up is incorrect and must be reworked.

Examples: (1) A question asks us to give the set-up for calculating speed in miles per hour and we arrive at the set-up:

$$\frac{120 \text{ miles}}{2 \text{ hours}} = \text{speed (miles/hour)}$$

A quick check tells us that we have a correct set-up with respect to units (but not necessarily with respect to numerical values). We have a miles unit divided by an hour unit on each side.

(2) Suppose a question asks us to give the set-up for a term called specific heat, which has the units of:

$$\frac{\text{calories}}{\text{grams x degrees}}$$

We have arrived at the following set-up:

$$\frac{50 \text{ calories}}{5 \text{ grams}} \text{ x } 20 \text{ grams x } \frac{1}{15 \text{ degrees}} = \text{specific heat}$$

A quick check shows that the gram units divide out, leaving us with calorie units divided by degree units,

$$\frac{\text{calories}}{\text{grams}} \text{ x grams x } \frac{1}{\text{degrees}} \neq \frac{\text{calories}}{\text{grams x degrees}}$$

Because the units in our set-up do not agree with the units required by the problem, we should rework our set-up.

Example (2) underlines an important concept in problem solving: *Always include units in the set-up and in all calculations.* A little self-discipline along these lines is definitely worth the effort.

We also want to be sure our numbers are "in the ball park." To estimate numbers, we look only at the numbers in the set-up, forgetting the units. We round off all numbers to one digit and zeroes. Then we write the rounded number in scientific notation. This way we can do the arithmetic in our heads. As an example, consider the following set-up.

$$\frac{1200 \text{ miles}}{20 \text{ hours}} = \text{speed (miles/hour)}$$

Concentrating only on the numbers, rounding off to one digit, and expressing in scientific notation we have:

$$\frac{1000}{20} = \frac{1 \times 10^3}{2 \times 10^1} = .5 \times 10^2 \text{ or } 5 \times 10^1 = 50$$

Now if our experience tells us that this answer is in the correct neighborhood for the problem, we can be satisfied that we have a good set-up. For example, if we are calculating the speed of an automobile, an answer of 50 is in the right range. However, if we were making a calculation on the speed of an interstellar spaceship (as in Star Trek) and we expected a speed of 5,000,000 miles per hour, then we should re-examine how we arrived at the numbers in our set-up.

PROBLEMS. Estimate the following.

(a) $\dfrac{1281 \times 813}{.00182}$ _____

(b) $\dfrac{100 \text{ miles}}{60 \text{ minutes}}$ = speed (mi/hr) _____

_ _ _ _ _ _ _ _ _ _ _ _ _ _ _ _

(a) $\dfrac{1000 \times 800}{.002} = \dfrac{(1 \times 10^3)(8 \times 10^2)}{2 \times 10^{-3}} = 4 \times 10^8$

(b) Cannot estimate since units do not check. Change 60 minutes to 1 hr. and estimate 100 mi/hr.

PRACTICE PROBLEMS

As promised, here are some problems to give you a chance to apply what you have learned in this chapter. After you have completed the problems, check your solutions with the answers that follow. If you wish to review any of these topics, reread the frames indicated in parentheses following the answers. Use a separate sheet of paper to write your answers.

1. **Use of Equations**
   Solve by rearranging the formula, cm = (2.54) in. How many inches in 15 cm?

2. **Dimensional analysis**
   Solve using dimensional analysis. Assume acceleration has the units, $\dfrac{\text{meters}}{(\text{second})(\text{second})}$. A car velocity increased 50 meters/second in 5 seconds. What was its acceleration?

3. **Metric Conversions**
   How many mm in 1.6 km?

4. **Conversions**
   Convert 8 quarts to liters.

5. **Concepts and Dimensional Analysis**
 (a) **Percent**
   In a certain population, 7% has blue eyes. If there are 357 blue-eyed people, what is the total population?

(b) **Density**
What is the volume of a substance that has a mass of 8 grams and a density of .32 g/cm^3?

(c) **Temperature Conversion**
313°K is how many °F?

(d) **Heat Change**
The specific heat of iron is 0.107 cal/g(deg). If a bar of iron increases 5°C by applying 10 calories of heat, what is the iron bar's mass?

(e) **Heat Transfer**
30 g of water at 27°C is mixed with 15 g of water at 40°C. What is the final temperature?

6. **Estimating**
Estimate $\dfrac{1825 \times (.009999)}{9.752}$

Answer to Practice Problems

1. in. $= \dfrac{cm}{2.54}$, so in. $= \dfrac{15}{2.54} = 5.91$ (frame 3)

2. 50 m/sec (1/5 sec) $= \dfrac{50\ m}{5\ (sec)(sec)} = \dfrac{10\ m}{(sec)(sec)}$ (frames 4-8)

3. 1000 m = 1 km, so 1 = 1000 m/km
    1.6 km (1000 m/km) = 1600 m
    1 m = 1000 mm, so 1 = 1000 mm/m
    1600 m (1000 mm/m) = 1.6 x 10^6 mm
    (frames 9-11)

4. $\dfrac{1\ liter}{1.06\ qt.} = 1$, so 8 qt. $\left(\dfrac{1\ liter}{1.06\ qt.}\right) = 7.55$ liters (frames 12-16)

5(a) 7% blue $= \dfrac{357\ blue}{pop} \times 100\%\ pop$

pop $= \dfrac{357\ blue \times 100\%\ pop}{7\%\ blue}$

pop = 5100 pop
(frames 17-19)

(b) $M = 8$ grams, $D = .32$ g/cm^3

$\dfrac{V}{D}\ D = \dfrac{M}{V}\cdot\dfrac{V}{D}$, so $V = \dfrac{M}{D}$

$V = \dfrac{8\ g}{.32}$ (cm^3/g)

$V = 25$ cm^3
(frames 20-23)

(c) $°K = °C + 273°$
$°K - 273° = °C$

Now, $°C = \dfrac{°F - 32°}{1.8}$, so substituting for $°C$ we get:

$(°K - 273°) = \dfrac{°F - 32°}{1.8}$. Multiplying both sides by 1.8 and rearranging, we get:

$1.8\,(°K - 273°) + 32° = °F$
$°K = 313°$ so we solve for $°F$:
$1.8\,(313° - 273°) + 32° = °F$
$104° = °F$

(frames 24-25)

(d) Heat change = (specific heat)(mass)(change in temperature)
Unknown = M (mass)
Heat change = +10 calories
Specific heat = 0.107 cal/g(deg)
Change in temp = +5°

$10 \text{ cal} = \dfrac{.107 \text{ cal}}{\text{g(deg)}} \cdot M \cdot 5 \text{ deg}$

$10 \text{ g} = .535\,M$

$\dfrac{10 \text{ g}}{.535} = M = 18.69 \text{ g}$

(frames 26-28)

(e) $T_{f-27°} = T_{f-40°}$; Heat gained by 27°water = Heat lost by 40° water. Specific heats cancel out since they are equal. Also we write $T_i - T_f$ so we don't get a negative for 40° side of equation.

$M_{27°}\,(T_f - T_i) = M_{40°}\,(T_i - T_f)$
$30\,(T_f - 27°) = 15(40° - T_f)$
$30T_f - 810 = 600 - 15T_f$
$45T_f = 1410$
$T_f = 31.3°C$

(frames 29-30)

6. $\dfrac{2000 \times (.01)}{10} = \dfrac{(2 \times 10^3) \times (1 \times 10^{-2})}{1 \times 10^1} = 2 \times 10^{3-2-1} = 2$

A Final Self-Test, covering all the material in the book, starts on page 136.

Final Self-Test

Chapter One

1. Carry out the indicated operations.

   (a) $-10 - (-8)$ _____

   (b) $a(b + 7)$ _____

   (c) $a(0)(1024)$ _____

   (d) $(-7)(-8)(-2)$ _____

   (e) $\dfrac{-18xy}{3x}$ _____

2. What are the factors in $(7)(8) = 56$? _____

3. What is the reciprocal of $\dfrac{8}{15}$? _____

4. What is the denominator in $\dfrac{112}{185}$? _____

Chapter Two

1. Carry out the indicated operations.

   (a) $\dfrac{a}{xy} - \dfrac{z}{xy}$ _____

   (b) $\dfrac{8}{y} - \dfrac{7}{x}$ _____

   (c) $\left(\dfrac{-13}{18}\right)\left(\dfrac{3}{26}\right)$ _____

   (d) $\dfrac{\dfrac{\text{fruit flies}}{\text{bottle}}}{\dfrac{\text{grams of food}}{\text{bottle}}}$ _____

(e) $\dfrac{\dfrac{3}{8}}{\dfrac{9}{16}}$ _____

(f) $\dfrac{\dfrac{1}{2} + \dfrac{1}{3}}{\dfrac{10}{11}}$ _____

Chapter Three

1. Indicate which sign (=, >, <) completes this statement.

   $3a + 2$ ____ 8, if $a < 2$

2. Express the following statement as an algebraic equation.

   The average test score is the sum of John's test score, Mary's test score, and Tom's test score divided by three.

   _____

3. From the equation above, find the average test score if the three students had the following test scores: John, 79; Mary, 98; Tom, 42.

   _____

4. Identify the independent and dependent variables in the equation from question 2 above. _____

5. Given: $x + 8 = 12$, solve for x. _____

6. Given: $\dfrac{A}{B} = \dfrac{C}{D}$, solve for C. _____

7. Solve the following equations for x.

   (a) $8(x + y) = 9y$ _____

   (b) $9x + 10 = y$ _____

   (c) $8x + xy = 10$ _____

8. One fourth of a freshman biology class has an IQ of over 125. An equal number of men and women in the class have an IQ of over 125. If three women have a 125+ IQ, how large is the biology class?

Chapter Four

1. Write the following, including constants, in terms of positive exponents.

$2ab\,3b\,2a\,5a\,3ab\,2a$ _____

2. Write the following in terms of negative exponents.

$\left(\dfrac{8}{9}\right)^4\left(\dfrac{1}{ab}\right)$ _____

3. Express the following first as a fractional exponent and then with a radical sign.

The xth root of $825b$ _____ .

4. Carry out the indicated operations and keep answers in exponential notation.

   (a) $\dfrac{x^5\,y^7\,x^8}{x^2\,y^{-5}}$ _____

   (b) $(\,(x+y)^6\,)^{\frac{1}{3}}$ _____

5. Solve for x.

   $y - 10 = x^7 - 4$ _____

6. Given: $x^2 - 4x = 5$, solve for x. _____

Chapter Five

1. Express .000756 in scientific notation. _____

2. Express 8.53×10^6 as a non-exponential number. _____

3. Carry out the indicated operations. (Use a separate sheet of paper for calculations if necessary.)

   (a) $8.0 \times 10^2 - 3.4 \times 10^4$ _____

   (b) $(3.1 \times 10^{-8})(2.0 \times 10^{10})$ _____

   (c) $\dfrac{1.21 \times 10^{-7}}{1.1 \times 10^{-9}}$ _____

(d) $(1.6 \times 10^{13})^{\frac{1}{4}}$ _____

(e) $(3.0 \times 10^{-1})^3$ _____

Chapter Six

1. Find the logs of the following numbers

    (a)   10,000 _____

    (b)   0.285 _____

    (c)   13,950 _____

2. Find the antilogs of the following logs.

    (a)   7.0 + 0.8414 _____

    (b)   −3.0 + 0.5000 _____

    (c)   −1.2993 _____

3. Carry out the indicated operations using logs.

    (a) $\dfrac{47.5 \times 779}{760}$

    (b) $\sqrt{643} \times (1.91)^3$

    (c) $(520.20)^{\frac{1}{3}}$

Answers to Self-Tests

Chapter One

1. (a)   −2
    (b) $ab + 7a$
    (c)   0
    (d)   −112
    (e)   −6y

2.   7 and 8

3. $\dfrac{15}{8}$

4.   185

Chapter Two

1. (a) $\dfrac{a - z}{xy}$

   (b) $\dfrac{8x - 7y}{xy}$

   (c) $\dfrac{-1}{12}$

   (d) $\dfrac{\text{fruit flies}}{\text{grams of food}}$

   (e) $\dfrac{2}{3}$

   (f) $\dfrac{11}{12}$

Chapter Three

1. $<$

2. $A = \dfrac{J + M + T}{3}$ (You may have chosen different letters, but the expression should be the same.)

3. 73

4. J, M, T = independent variables
    A = dependent variable

5. 4

6. $C = \dfrac{AD}{B}$

7. (a) $x = \dfrac{y}{8}$

   (b) $x = \dfrac{y - 10}{9}$

   (c) $x = \dfrac{10}{8 + y}$

8. 24

Chapter Four

1. $2^3\, 3^2\, 5^1\, a^5\, b^3$

2. $\left(\dfrac{9}{8}\right)^{-4} a^{-1} b^{-1}$

3. $(825b)^{\frac{1}{x}}$; $\sqrt[x]{825b}$

4. (a) $x^{11}y^{12}$
   (b) $(x + y)^2$

5. $(y - 6)^{\frac{1}{7}}$

6.   5, −1

Chapter Five

1. 7.56×10^{-4}

2.   8,530,000

3. (a) -3.32×10^4
   (b) 6.2×10^2
   (c) 1.1×10^2
   (d) 2×10^3
   (e) 2.7×10^{-2}

Chapter Six

1. (a) 4
   (b) −1.0 + 0.4548
   (c) 4.0 + 0.1446

2. (a) 6.94×10^7
   (b) 3.16×10^{-3}
   (c) 5.02×10^{-2}

3. (a) 48.7
   (b) 66.9
   (c) 8.04

Appendix

COMMON UNITS OF WEIGHTS AND MEASURES

Long Measure

12 inches	= 1 foot
3 feet or 36 inches	= 1 yard
$5\frac{1}{2}$ yards	= 1 rod
1760 yards or 5280 feet	= 1 mile

Square Measure

144 square inches (12 x 12)	= 1 square foot
9 square feet (3 x 3)	= 1 square yard
$30\frac{1}{4}$ square yards ($5\frac{1}{2}$ x $5\frac{1}{2}$)	= 1 square rod
160 square rods or 43,560 square feet	= 1 acre
640 acres	= 1 square mile

Cubic Measure

1723 cubic inches (12 x 12 x 12)	= 1 cubic foot
27 cubic feet (3 x 3 x 3)	= 1 cubic yard

Ordinary Weight

16 ounces	= 1 pound
100 pounds	= 1 hundredweight
2000 pounds	= 1 short ton
2240 pounds	= 1 long ton

Common Household Measure

1 cup	= 8 ounces, liquid
2 cups (8 ounces each)	= 1 pint or 16 ounces, liquid measure
2 pints (16 ounces each)	= 1 quart, liquid or 4 cups
4 quarts, liquid	= 1 gallon
8 quarts, dry	= 1 peck
4 pecks or 32 quarts	= 1 bushel

COMMON LOGARITHMS

Logarithms   Proportional Parts

No.	0	1	2	3	4	5	6	7	8	9	1	2	3	4	5	6	7	8	9
10	0000	0043	0086	0128	0170	0212	0253	0294	0334	0374	4	8	12	17	21	25	29	33	37
11	0414	0453	0492	0531	0569	0607	0645	0682	0719	0755	4	8	11	15	19	23	26	30	34
12	0792	0828	0864	0899	0934	0969	1004	1038	1072	1106	3	7	10	14	17	21	24	28	31
13	1139	1173	1206	1239	1271	1303	1335	1367	1399	1430	3	6	10	13	16	19	23	26	29
14	1461	1492	1523	1553	1584	1614	1644	1673	1703	1732	3	6	9	12	15	18	21	24	27
15	1761	1790	1818	1847	1875	1903	1931	1959	1987	2014	3	6	8	11	14	17	20	22	25
16	2041	2068	2095	2122	2148	2175	2201	2227	2253	2279	3	5	8	11	13	16	18	21	24
17	2304	2330	2355	2380	2405	2430	2455	2480	2504	2529	2	5	7	10	12	15	17	20	22
18	2553	2577	2601	2625	2648	2672	2695	2718	2742	2765	2	5	7	9	12	14	16	19	21
19	2788	2810	2833	2856	2878	2900	2923	2945	2967	2989	2	4	7	9	11	13	16	18	20
20	3010	3032	3054	3075	3096	3118	3139	3160	3181	3201	2	4	6	8	11	13	15	17	19
21	3222	3243	3263	3284	3304	3324	3345	3365	3385	3404	2	4	6	8	10	12	14	16	18
22	3424	3444	3464	3483	3502	3522	3541	3560	3579	3598	2	4	6	8	10	12	14	15	17
23	3617	3636	3655	3674	3692	3711	3729	3747	3766	3784	2	4	6	7	9	11	13	15	17
24	3802	3820	3838	3856	3874	3892	3909	3927	3945	3962	2	4	5	7	9	11	12	14	16
25	3979	3997	4014	4031	4048	4065	4082	4099	4116	4133	2	3	5	7	9	10	12	14	15
26	4150	4166	4183	4200	4216	4232	4249	4265	4281	4298	2	3	5	7	8	10	11	13	15
27	4314	4330	4346	4362	4378	4393	4409	4425	4440	4456	2	3	5	6	8	9	11	13	14
28	4472	4487	4502	4518	4533	4548	4564	4579	4594	4609	2	3	5	6	8	9	11	12	14
29	4624	4639	4654	4669	4683	4698	4713	4728	4742	4757	1	3	4	6	7	9	10	12	13
30	4771	4786	4800	4814	4829	4843	4857	4871	4886	4900	1	3	4	6	7	9	10	11	13
31	4914	4928	4942	4955	4969	4983	4997	5011	5024	5038	1	3	4	6	7	8	10	11	12
32	5051	5065	5079	5092	5105	5119	5132	5145	5159	5172	1	3	4	5	7	8	9	11	12
33	5185	5198	5211	5224	5237	5250	5263	5276	5289	5302	1	3	4	5	6	8	9	10	12
34	5315	5328	5340	5353	5366	5378	5391	5403	5416	5428	1	3	4	5	6	8	9	10	11
35	5441	5453	5465	5478	5490	5502	5514	5527	5539	5551	1	2	4	5	6	7	9	10	11
36	5563	5575	5587	5599	5611	5623	5635	5647	5658	5670	1	2	4	5	6	7	8	10	11
37	5682	5694	5705	5717	5729	5740	5752	5763	5775	5786	1	2	3	5	6	7	8	9	10
38	5798	5809	5821	5832	5843	5855	5866	5877	5888	5899	1	2	3	5	6	7	8	9	10
39	5911	5922	5933	5944	5955	5966	5977	5988	5999	6010	1	2	3	4	5	7	8	9	10
40	6021	6031	6042	6053	6064	6075	6085	6096	6107	6117	1	2	3	4	5	6	8	9	10
41	6128	6138	6149	6160	6170	6180	6191	6201	6212	6222	1	2	3	4	5	6	7	8	9
42	6232	6243	6253	6263	6274	6284	6294	6304	6314	6325	1	2	3	4	5	6	7	8	9
43	6335	6345	6355	6365	6375	6386	6395	6405	6415	6425	1	2	3	4	5	6	7	8	9
44	6435	6444	6454	6464	6474	6484	6493	6503	6513	6522	1	2	3	4	5	6	7	8	9
45	6532	6542	6551	6561	6571	6580	6590	6599	6609	6618	1	2	3	4	5	6	7	8	9
46	6628	6637	6646	6656	6665	6675	6684	6693	6702	6712	1	2	3	4	5	6	7	7	8
47	6721	6730	6739	6749	6758	6767	6776	6785	6794	6803	1	2	3	4	5	5	6	7	8
48	6812	6821	6830	6839	6848	6857	6866	6875	6884	6893	1	2	3	4	4	5	6	7	8
49	6902	6911	6920	6928	6937	6946	6955	6964	6972	6981	1	2	3	4	4	5	6	7	8
50	6990	6998	7007	7016	7024	7033	7042	7050	7059	7067	1	2	3	3	4	5	6	7	8
51	7076	7084	7093	7101	7110	7118	7126	7135	7143	7152	1	2	3	3	4	5	6	7	8
52	7160	7168	7177	7185	7193	7202	7210	7218	7226	7235	1	2	2	3	4	5	6	7	7
53	7243	7251	7259	7267	7275	7284	7292	7300	7308	7316	1	2	2	3	4	5	6	6	7
54	7324	7332	7340	7348	7356	7364	7372	7380	7388	7396	1	2	2	3	4	5	6	6	7
	0	1	2	3	4	5	6	7	8	9	1	2	3	4	5	6	7	8	9

COMMON LOGARITHMS (continued)

Logarithms   Proportional Parts

No.	0	1	2	3	4	5	6	7	8	9	1 2 3	4 5 6	7 8 9
55	7404	7412	7419	7427	7435	7443	7451	7459	7466	7474	1 2 2	3 4 5	5 6 7
56	7482	7490	7497	7505	7513	7520	7528	7536	7543	7551	1 2 2	3 4 5	5 6 7
57	7559	7566	7574	7582	7589	7597	7604	7612	7619	7627	1 2 2	3 4 5	5 6 7
58	7634	7642	7649	7657	7664	7672	7679	7686	7694	7701	1 1 2	3 4 4	5 6 7
59	7709	7716	7723	7731	7738	7745	7752	7760	7767	7774	1 1 2	3 4 4	5 6 7
60	7782	7789	7796	7803	7810	7818	7825	7832	7839	7846	1 1 2	3 4 4	5 6 6
61	7853	7860	7868	7875	7882	7889	7896	7903	7910	7917	1 1 2	3 4 4	5 6 6
62	7924	7931	7938	7945	7952	7959	7966	7973	7980	7987	1 1 2	3 3 4	5 6 6
63	7992	8000	8007	8014	8021	8028	8035	8041	8048	8055	1 1 2	3 3 4	5 5 6
64	8062	8069	8075	8082	8089	8096	8102	8109	8116	8122	1 1 2	3 3 4	5 5 6
65	8129	8136	8142	8149	8156	8162	8169	8176	8182	8189	1 1 2	3 3 4	5 5 6
66	8195	8202	8209	8215	8222	8228	8235	8241	8248	8254	1 1 2	3 3 4	5 5 6
67	8261	8267	8274	8280	8287	8293	8299	8306	8312	8319	1 1 2	3 3 4	5 5 6
68	8325	8331	8338	8344	8351	8357	8363	8370	8376	8382	1 1 2	3 3 4	4 5 6
69	8388	8395	8401	8407	8414	8420	8426	8432	8439	8445	1 1 2	2 3 4	4 5 6
70	8451	8457	8463	8470	8476	8482	8488	8494	8500	8506	1 1 2	2 3 4	4 5 6
71	8513	8519	8525	8531	8537	8543	8549	8555	8561	8567	1 1 2	2 3 4	4 5 5
72	8573	8579	8585	8591	8597	8603	8609	8615	8621	8627	1 1 2	2 3 4	4 5 5
73	8633	8639	8645	8651	8657	8663	8669	8675	8681	8686	1 1 2	2 3 4	4 5 5
74	8692	8698	8704	8710	8716	8722	8727	8733	8739	8745	1 1 2	2 3 4	4 5 5
75	8751	8756	8762	8768	8774	8779	8785	8791	8797	8802	1 1 2	2 3 3	4 5 5
76	8808	8814	8820	8825	8831	8837	8842	8848	8854	8859	1 1 2	2 3 3	4 5 5
77	8865	8871	8876	8882	8887	8893	8899	8904	8910	8915	1 1 2	2 3 3	4 4 5
78	8921	8927	8932	8938	8943	8949	8954	8960	8965	8971	1 1 2	2 3 3	4 4 5
79	8976	8982	8987	8993	8998	9004	9009	9015	9020	9025	1 1 2	2 3 3	4 4 5
80	9031	9036	9042	9047	9053	9058	9063	9069	9074	9079	1 1 2	2 3 3	4 4 5
81	9085	9090	9096	9101	9106	9112	9117	9122	9128	9133	1 1 2	2 3 3	4 4 5
82	9138	9143	9149	9154	9159	9165	9170	9175	9180	9186	1 1 2	2 3 3	4 4 5
83	9191	9196	9201	9206	9212	9217	9222	9227	9232	9238	1 1 2	2 3 3	4 4 5
84	9243	9248	9253	9258	9263	9269	9274	9279	9284	9289	1 1 2	2 3 3	4 4 5
85	9294	9299	9304	9309	9315	9320	9325	9330	9335	9340	1 1 2	2 3 3	4 4 5
86	9345	9350	9355	9360	9365	9370	9375	9380	9385	9390	1 1 2	2 3 3	4 4 5
87	9395	9400	9405	9410	9415	9420	9425	9430	9435	9440	0 1 1	2 2 3	3 4 4
88	9445	9450	9455	9460	9465	9469	9474	9479	9484	9489	0 1 1	2 2 3	3 4 4
89	9494	9499	9504	9509	9513	9518	9523	9528	9533	9538	0 1 1	2 2 3	3 4 4
90	9542	9547	9552	9557	9562	9566	9571	9576	9581	9586	0 1 1	2 2 3	3 4 4
91	9590	9595	9600	9605	9609	9614	9619	9624	9628	9633	0 1 1	2 2 3	3 4 4
92	9638	9643	9647	9652	9657	9661	9666	9671	9675	9680	0 1 1	2 2 3	3 4 4
93	9685	9689	9694	9699	9703	9708	9713	9717	9722	9727	0 1 1	2 2 3	3 4 4
94	9731	9736	9741	9745	9750	9754	9759	9763	9768	9773	0 1 1	2 2 3	3 4 4
95	9777	9782	9786	9791	9795	9800	9805	9809	9814	9818	0 1 1	2 2 3	3 4 4
96	9823	9827	9832	9836	9841	9845	9850	9854	9859	9863	0 1 1	2 2 3	3 4 4
97	9868	9872	9877	9881	9886	9890	9894	9899	9903	9908	0 1 1	2 2 3	3 4 4
98	9912	9917	9921	9926	9930	9934	9939	9943	9948	9952	0 1 1	2 2 3	3 4 4
99	9956	9961	9965	9969	9974	9978	9983	9987	9991	9996	0 1 1	2 2 3	3 3 4
	0	1	2	3	4	5	6	7	8	9	1 2 3	4 5 6	7 8 9

Index

Addition and Subtraction, 18-22, 72-73
Angstrom, 113
Antilogs, 88-91, 100-101
Applications, 46-50, 63-64

Calorie, 124
Centigrade, 123
Characteristic, 85
Common denominator, 18
Complex fractions, 26
Constants, 35
Conversion factors, 109
Conversions, 71-72, 115-117, 123-124

Decimal fraction, 10
Density, 120-123
Dependent variable, 37
Dimensional analysis, 108-112
Distributing, 7
Dividend, 9
Dividing out, 11
Division, 25-26, 74, 91
Divisor, 8

Equations, 32-34, 107-108
Estimation, 131-133
Exponents, 55-63, 94
Expressing numbers in scientific notation, 70-71

Factor, 6
Factoring, 45
Fahrenheit, 123
Fractions, 9, 18-22
Fractional exponents, 58-60

Gram, 113

Heat change, 124-128
Heat transfer, 128-131

Improper fraction, 21
Independent variable, 37
Inverse, 10

Kelvin, 123

Liter, 123
Logarithms, 80-84
    division, 91
    exponent rule, 94
    multiplication rule, 91
    root rule, 95
    tables, 84-86, 144-145

Mantissa, 85
Mass, 120
Meter, 112
Metric conversions, 112-115
Mixed number, 22
Multiplication, 24, 73-74, 91

Negative characteristic, 96
Negative exponents, 57-58
Negative logarithms, 100-101

Operations on exponents, 60-63
Order of operations, 76

p functions, 101-104
Per, 10
Percentage, 118-120
Power, 55
Problem solving, 107-112
Product, 6
Proportional parts table, 86-88, 90-91

Quadratic formula, 64-67
Quotient, 9

Radical, 58
Raising to a power, 75-76
Ratio, 9

Reciprocal, 10
Root, 58
Root rule, 95

Scientific notation, 70-76
Signs, 4-5, 7-8, 13-14
Solving equations, 36-38
Specific heat, 124

Standard unit, 112
Substitution, 35-36

Taking roots, 75
Temperature conversion, 123-124

Variables, 35, 37
Volume, 120